The Teacher's Guide to Scratch – Intermediate

The Teacher's Guide to Scratch – Intermediate is a practical guide for educators preparing moderately complex coding lessons and assignments in their K–12 classrooms. The world's largest and most active visual programming platform, Scratch helps today's schools answer the growing call to realize important learning outcomes using coding and computer science. This book illustrates the increasingly intricate affordances of Scratch coding, details effective pedagogical strategies and learner collaborations, and offers actionable, accessible troubleshooting tips. Geared toward the intermediate user, these four unique coding projects will provide the technical training that teachers need to feel comfortable and confident in their skills and to help instil the same feeling of accomplishment in their students. Clear goals, a comprehensive glossary, and other features ensure the project's enduring relevance as a reference work for computer science education in grade school. Thanks to Scratch's cost-effective open-source license, suitability for blended and project-based learning, notable lack of privacy or security risks, and consistency in format even amid software and interface updates, this will be an enduring practitioner manual and professional development resource for years to come.

Kai Hutchence is CEO and Founder of Massive Corporation Game Studios as well as its subdivision, Massive Learning, which focuses on educational products and services. Through his established coding support partnerships with elementary, middle and high schools, post-secondary institutions, and provincial and national organizations, Kai has taught over 20,000 students to code and over 2,000 educators to code and teach coding.

The Teacher's Guide to Scratch – Intermediate

Professional Development for Coding Education

Kai Hutchence

NEW YORK AND LONDON

Designed cover image: © Shutterstock

First published 2024
by Routledge
605 Third Avenue, New York, NY 10158

and by Routledge
4 Park Square, Milton Park, Abingdon, Oxon, OX14 4RN

Routledge is an imprint of the Taylor & Francis Group, an informa business

© 2024 Kai Hutchence

The right of Kai Hutchence to be identified as author of this work has been asserted in accordance with sections 77 and 78 of the Copyright, Designs and Patents Act 1988.

All rights reserved. No part of this book may be reprinted or reproduced or utilised in any form or by any electronic, mechanical, or other means, now known or hereafter invented, including photocopying and recording, or in any information storage or retrieval system, without permission in writing from the publishers.

Trademark notice: Product or corporate names may be trademarks or registered trademarks, and are used only for identification and explanation without intent to infringe.

ISBN: 978-1-032-49911-6 (hbk)
ISBN: 978-1-032-50566-4 (pbk)
ISBN: 978-1-003-39907-0 (ebk)

DOI: 10.4324/9781003399070

Typeset in Palatino
by Apex CoVantage, LLC

Additional copyediting and developmental editing provided by
Gui de Souza Rocha.

Scratch is a project of the Scratch Foundation, in collaboration with the Lifelong Kindergarten Group at the MIT Media Lab. It is available for free at https://scratch.mit.edu.

The Scratch name, Scratch logo, Scratch Day logo, Scratch Cat, and Gobo are Trademarks owned by the Scratch Team and are used for identification and do not constitute or imply ownership or endorsement by the Scratch Foundation or Lifelong Kindergarten Group at the MIT Media Lab.

With thanks to Gui de Souza Rocha and Terry Hoganson for their assistance, encouragement, editing, testing, and feedback.

Contents

Meet the Author	*xii*
Foreword	*xiii*
Reader's Key	*xv*
1 Introduction	**1**
2 Our Previous Book in the Series	**4**
3 Scratch's Place in Education: General Education or a Specialized Field?	**6**
4 Defining Intermediate Scratch	**8**
Intermediate Scratch Attitudes	8
Intermediate Scratch Goals	10
5 Intermediate Project 1: Pen Tool Fun	**11**
What This Project Is	11
What We're Learning with It	12
Building It	12
Step 0: Create Your New Project	12
Step 1: Adding the Pen Extension	12
Step 2: Our First Drawing	13
Step 3: Reset	14
Step 4: Our First Shape	15
Step 5: Our Second Shape	16
Step 6: A Complication	17
Step 7: Variable Sides	18
Step 8: Shapes Sequence	20
Step 9: Nested Shape Patterns	21
Step 10: Randomizing Function	23
Step 11: Stamping Patterns	23
Step 12: Grid Terrain Generation	26
Step 13: Customization	29
6 Intermediate Project 2: Interactive Story	**32**
What This Project Is	32
What We're Learning with It	32

    Building It   34
        Step 0: Create Your New Project   34
        Step 1: Goblin Storyteller   34
        Step 2: Messaging Events   35
        Step 3: Wandering   36
        Step 4: Handling Human Answers   36
        Step 5: Personalizing Things   38
        Step 6: Our First Scene Objects   39
        Step 7: The Wizard Sets the Story   41
        Step 8: A Fork in the Path   43
        Step 9: Off to the Farm   45
        Step 10: Into the Desert   48
        Step 11: Into the Woods   48
        Step 12: The Gate and the Ogre   51
        Step 13: Feeding the Ogre   52
        Step 14: Testing Potions   55
        Step 15: Potion #1 – Balloon Transformation   58
        Step 16: Potion #2 – Moustache Growth   62
        Step 17: Potion #3 – Ghost Transformation   63
        Step 18: Win Screen   66

7 Intermediate Project 3: Snowball Fight   69
    What This Project Is   69
    What We're Learning with It   69
    Building It   70
        Step 0: Create Your New Project   70
        Step 1: An Arcing Snowball   71
        Step 2: Throwing Angle   72
        Step 3: Throwing Power   73
        Step 4: Aiming   74
        Step 5: Throw Event   76
        Step 6: Our Target   76
        Step 7: The Player   78
        Step 8: The Turn Sign   80
        Step 9: Player Turns   81
        Step 10: The Snowman Strikes Back!   83
        Step 11: Improved Controls   85
        Step 12: A Palpable Hit!   87
        Step 13: An Extra Challenge   88
        Step 14: The GameStart Event   91
        Step 15: Wind Effects   94
        Step 16: Bonus Stars   96

	Step 17: Defeat	98
	Step 18: Start Screen	100

8 Intermediate Project 4: Big Map Racing — 102
What This Project Is — 102
What We're Learning with It — 103
Building It — 104
    Step 0: Create Your New Project — 104
    Step 1: Drawing a Racetrack — 104
    Step 2: Drawing a Car — 106
    Step 3: Turning the Car — 106
    Step 4: Moving the Car/Map — 109
    Step 5: Off-Roading — 111
    Step 6: Waypoints and Lap Counting — 112
    Step 7: Adding a Timer — 113
    Step 8: Driving and Racing States — 115
    Step 9: Ending the Race — 116
    Step 10: Barriers and Crashing — 117
    Step 11: Countdown — 121
    Step 12: Nitro Boosts — 123
    Step 13: The Main Menu Screen — 124
    Step 14: Menu Map Display — 128
    Step 15: Multiple Racetracks — 129

9 Intermediate Check-In — 133
Key Skills — 133
    Events and Triggers — 134
        Messaging — 134
        Custom Blocks — 135
        Until Loops — 135
    Dealing with Scratch Limits — 135
        Size Limits — 135
        Position Limits — 136
    Variables — 136
        Math Altering — 137
        Timers — 138
        Randomizing — 138
        Limiting, or Clamping — 138
        Economy or Usage Limits — 138
        Hiding or Showing — 138
        Object Referencing — 138

Inputs	138
Key Press Tests	139
Text Input and Saving	139
Colour Collisions	139
Logic Connectors	140
Game States/State Machines	140
Game Turns	140
Control Locking	140
Levels and Tracks	140
Event and Stage Switching	141
Building Interfaces and Menus	142
Messaging Calls	142
Layering Elements	142
Variable Setting Buttons	143
Clones	143
Temporary Objects	143
Multiple Objects	144
Original vs. Clone	144
More Intermediate Practice	144
Infinite Swimmer	145
Interactive Diorama	145
Pong	145
Teaching Intermediate Scratch	146
Working with Scratch	146
Logic	148
Project Design	150

10 Follow-Up: Extending the Projects — 151

Commenting	152
Dinosaur Dance Party	153
Fireworks Display	153
Batty Flaps	153
Butterfly Catcher	154
Pen Tool Fun	154
Interactive Story	154
Snowball Fight	155
Big Map Racing	155
Advanced Scratch	155

11 Troubleshooting Scratch — 157

Site Issues	158

	Coding Issues	159
	The Wrong Object	160
	General Tips	161
	The Wrong Block	162
	Confused Pair: Left vs. Right	162
	Confused Pair: Go To vs. Glide To	162
	Confused Pair: X vs. Y	163
	Confused Pair: Set vs. Change	164
	Confused Pair: Say/Think vs. Say/Think For	164
	Confused Pair: Play Sound vs. Start Sound	164
	Confused Pair: If vs. If/Else	165
	Confused Pair: > vs. <	165
	The Wrong Order	165
	Simple Sequences Code Flow	165
	Control Structures and Code Flow	166
	Logic Clauses and Chains	166
	Concurrency and Race Conditions	167
	Other Errors	167
	Layers	167
	Visibility	168
	Colour Selection	168
	Clones vs. Originals	168
	Wrong Concepts	168
	Backup Plans	169
	Offline Scratch	169
	Pseudo-Coding	170
12	**The Next Step in Your Coding Journey**	**171**
13	**Final Thoughts**	**173**
	Glossary	*175*
	Index	*193*

Meet the Author

Born and raised in Regina, Saskatchewan, Kai Hutchence was fortunate to be the son of a math professor (and later research scientist) with an interest in computer science. With his father's help, he taught himself BASIC coding and, as a teenager, HTML, Visual Basic, and other languages. Living in a province with, at that time, almost no tech companies, Kai explored careers in politics, non-profits, and restaurateuring before moving to Ontario to work in game development.

In Ontario, he co-founded a game studio that went on to garner over four million downloads of its mobile apps. Living amidst a thriving game development community, he made connections with major players in the industry, helped aspiring creators publish their own games, and saw a community transform into an industry. After almost a decade in Ontario, he moved back to Saskatchewan to take those lessons and try to build up the industry in his home province, a lifelong goal.

In Saskatchewan, Kai launched SaskGameDev as an organization to build, nurture, and support game development in the province. Additionally, he launched his own game development company, Massive Corporation Game Studios. To support the long-term growth of game development in the province, he took up a consultancy role to help spread coding education through schools and non-profits.

While dedicating a portion of his time to teaching, Kai has helped establish coding support partnerships with elementary, middle and high schools, post-secondary institutions, and provincial and national organizations. He has helped develop coding and AI instructional material for nationwide use. Through his multiple partnerships and public profile, he has taught workshops, given lectures, provided industry mentorship and vouching, and facilitated internships with post-secondary institutes across Canada. Following up this success and newfound passion for coding education, Kai launched Massive Learning, a subdivision of Massive Corporation, to focus on educational products and services.

He currently lives in Regina, Saskatchewan. He continues to make games through Massive Corporation and provide educational services though Massive Learning. He enjoys cooking, gardening, and hiking, when he's not teaching, writing, or developing.

Foreword
by Karen Brennan

As a member of the MIT Scratch Team in the early days of Scratch, I received many emails from educators who were excited by the wide range of projects that they were seeing young people create in the online community. But they weren't entirely sure how these beautiful and inspiring projects might translate into activities for their own learning contexts and for the learners they supported. They were eager for ideas and resources.

For more than 15 years since Scratch's launch in 2007, I've been working with educators to help them include Scratch in a variety of contexts, from classrooms to libraries, to museums, and beyond. Scratch educators are a universally passionate, enthusiastic, and generous group. I have been particularly appreciative of and inspired by educators who have incorporated creative project activities into their learning environments. And I have been especially grateful to those educators who have documented their approaches and shared their ideas with others.

Over this time, an enormous number of resources have been developed to help more young people experience creative learning with Scratch. As with any type of learning support, no structure works for every learner – and so new approaches and resources are always needed. Kai Hutchence's book series is a much-welcomed addition to this ecosystem of support.

Though each resource is a special contribution in its own way, I want to offer three specific appreciations for what Kai shares with the world in this book: the importance of educators, the importance of intermediate experiences, and the importance of projects.

First, the importance of educators. Educators are too often written out of creative computing education experiences. Sometimes it's due to a perception that they lack the requisite content knowledge; sometimes it's due to a sense that educators should get out of the way of learner exploration. But we know from both educational research and practice that educators can offer invaluable support for creative learning. Kai Hutchence underscores the importance of the educator in the learning environment, helping learners through the cognitive and affective challenges they will encounter in their work. He offers a beautiful vision of the educator as a model learner – a member of the learning community who can be curious and vulnerable and fearless and brave.

Second, the importance of intermediate experiences. There are many beautiful, introductory, "first project" resources available in the world, both for Scratch and for many other programming environments for young learners. These foundational resources are so important; we know that bad first experiences can lead to prolonged discouragement. But what happens next? Even after a supportive initial experience, many learners still lack the capacity to create whatever they want on their own. As such, there is a need for "intermediate" learning resources – resources that acknowledge a learner's initial experiences, building on existing knowledge and offering support to go further in their learning. Kai's book responds to this need for deeper experiences.

Finally, the importance of projects. As a discipline, computer science has suffered from a concept-first approach: *first* help learners become acquainted with key computational concepts, and *then* create opportunities for them to put the concepts into practice through meaningful activities and projects. The challenge with this approach, of course, is that it takes too long to see the beauty and power of computing. Why would learners want to persist through challenges that don't matter to them? Kai orients us instead to a project-centric approach. By *starting* with projects, learners get experience and explore concepts while simultaneously developing creative work that is interesting and exciting to them. In this book, you will encounter powerful computational concepts, situated in projects that cut across a range of genres, including art, stories, games, and more.

With Kai, you have found a passionate and enthusiastic guide who is invested in your creative success. I can't wait to see how this book might help you – and the learners you support – to bring your wonderful, creative ideas into the world through coding.

Karen Brennan is Timothy E. Wirth Professor of Practice in Learning Technologies at Harvard Graduate School of Education, USA. She directs the Creative Computing Lab, co-chairs the Learning Design, Innovation, and Technology master's program, and serves as an affiliate of the Computer Science department. Her research focuses on the design of K–12 computer science learning experiences, particularly emphasizing self-direction within supportive learning communities. Prior to joining Harvard, Brennan earned her PhD at the MIT Media Lab, where she was a member of the team that developed the Scratch programming environment.

Reader's Key

Sidebars

Extensions – expanding Scratch's power	13
Over the Edge – Scratch position limitations	17
Geometry – 360 degrees in every shape	19
Testing Text – language and specificity	38
Functions – Using custom My Blocks	60
Solutions to Fit – Different solutions to the same problem	72
Out of Bounds – Keeping numbers where you want them	85
Colour Detection – Caution and confusion with colours	112

Style Legend

To help keep the different concepts involved clear for readers, we have adopted the following text stylings to denote particular things relating to Scratch projects:

Style – meaning
Object – a sprite or the stage
•Code Category – one of the colour-coded categories of code blocks
[Code Block] or **(Code Block)** or **<Code Block>** – any one of the many code components in Scratch (the brackets help convey the shape of the code block)
"Variable Name" – a variable added to the project
//Script Name – a connected sequence of code blocks (a stack or script) in a project object

1

Introduction

Welcome to Book 2 of *The Teacher's Guide to Scratch* series! This series was developed as your all-in-one guide to becoming proficient with coding in Scratch so you'll be ready to bring it into your classroom practice. We've covered off an introductory beginner's course in *The Teacher's Guide to Scratch – Beginner: Professional Development for Coding Education*, our first book, so hopefully you're ready to step things up with our intermediate projects!

If you haven't used Scratch before or are looking to just get started with coding, we'd highly recommend starting with our first book in the series. In it we provide an excellent and easy start to working with Scratch. We give you a complete guide to all of Scratch's capabilities that will help you understand where to find everything and how to work with its various tools. This book takes things to the next level. It won't go so thoroughly through the basics of finding code blocks, how to work with the different tabs, or other fundamentals. This book assumes you're comfortable with the editor and are ready to take the next leap. You'll notice our instructions are more streamlined in this book, having dropped the low-level detail of finding code blocks, and instead can provide more contextual information about the techniques rather than the technicalities. If you aren't there yet, check out Book 1 and get that solid foundation of familiarity with Scratch. It won't just be homework; the four beginner projects are fun and interesting and provide some really useful template projects to explore with students while teaching the core skills. Even if you are teaching higher grades, if you need to deal with remedial students or playing catch-up with some students, those simpler projects can be a good way for them to learn core skills to help them catch up, rather than letting them flounder attempting to understand more complex projects that make things a lot harder to follow.

Now, what are we doing in this book? You can look forward to seeing how projects in Scratch scale up. Compared to our first four projects in Book 1, the four projects in this book are larger and more complex. We're going to get into making multi-scene stories and multi-level games, some with multiple objects interacting. We're still focused on making projects you can adapt; you'll find many of the projects act as templates that can be re-imagined for other stories or themes, or just customized into personal explorations of a genre.

Figure 1.1 The four intermediate projects we cover in this book, the second in The Teacher's Guide to Scratch series.

The four intermediate-level projects are suitable for middle grades (approximately grades 5 to 8). Moving past just trying to familiarize you with Scratch, they introduce more complex concepts and methods in coding. The projects are considerably larger in scope, likely taking multiple sessions in class to complete. In Pen Tool Fun, we'll show how you can use programming to draw out geometric principles and patterns. Our Interactive Story project will introduce working with multiple scenes and inventory or story switch systems. The Snowball Fight game will explore more complex movement systems, give an example of enemy turns, and show randomization techniques for games. The Big Map Racing will reveal some handy techniques for working around some Scratch limitations, some handy progress tracking, and even how you can make and use multiple levels (racetracks in this case) in a game. Again, at the end of the projects, there is a check-in chapter to review what you've learned, give suggestions on teaching intermediate Scratch, as well as suggest additional intermediate project ideas.

We'll follow up our four projects and check-in chapters with some follow-up ideas on how you can challenge yourself and your students by adding tweaks and features to the projects in both Books 1 and 2 with all the skills you've learned over the course of both books. These are great ways to handle your coding whizz-kid students to keep them challenged but still on the same projects as everyone else. Then we'll give a rundown on the most common problems faced in Scratch and a glossary of terms. These are the same as in Book 1, but we wanted to make sure nobody missed out if they jumped in later in the series, because the ability to solve problems and accurately describe things are universal needs.

2
Our Previous Book in the Series

Before we get started with the new, intermediate projects, let's take a moment to reflect on the Book 1 projects. If you've already done them, then you'll be able to clue in to the techniques we revisit or expand on, but if you haven't, let's just cover the projects we did in brief and refresh some of the skills we expect readers to have to move forward with the new projects.

Our previous book, *The Teacher's Guide to Scratch – Beginner: Professional Development for Coding Education*, was all about introducing Scratch. It served to familiarize readers with all the amazing features in Scratch and provide a reference document for it. In the Code tab, we showed how to work with code blocks, the meaning behind their shapes and colours, how to find them, how to connect them, and how to alter or edit them with values and other code blocks. We showed off the amazing Costumes (or Backdrops) tab and the wonderful built-in digital art program in Scratch. We introduced the various tools, and we put them to use making new costumes or editing the built-in ones from the Sprite Library. We checked out all the sound effects and music loops in the Sounds tab and showed how to play background music and reactive music and sound effects. A full section of the book detailed out all the components in the Scratch editor.

Figure 2.1 The four projects covered in our first book in the series: The Teacher's Guide to Scratch – Beginners: Professional Development for Coding Education

The four projects in the book gave users a diverse sampling of the kinds of projects Scratch can make. In the Dino Dance Party project, we showed the basics of animation and playing music with some basic repeat loops, wait blocks for timing, and clickable sprites. The Fireworks Display project taught readers about the importance of a reset function while providing a template for animated artistic creativity. It explored the •Looks blocks to show, hide, and resize objects as well as use special graphic effects and the art tools to make new sprites and edit existing backdrops. In Batty Flaps, we created a clone of the smash-hit Flappy Bird game. We learned to use messages to create some menu functionality that allowed convenient replay of the game as well as more techniques for game controls and conditions. Lastly, in the Butterfly Catcher project, we made a game to show off variables, game states, and conditionals, so the game could end and present players with commentary on their scores.

With those projects under your belt, you should be comfortable with basic motion, keyboard, and mouse controls. We'd expect a basic familiarity with variables and conditionals in Scratch, but we'll be teaching a lot more about them in this book. You've hopefully worked with the sounds and made some custom art in the Costumes or Backdrops tab. If you've done a few projects in Scratch and you know your way around but are curious for more, let's go ahead and get learning about intermediate Scratch!

3

Scratch's Place in Education: General Education or a Specialized Field?

Computer science has been on offer in high schools for a generation or two in almost every jurisdiction in North America. It's always been handled as an elective subject, a specialized science, just like biology or chemistry. Why does this need to change?

For one, there's an issue of scale. The tech industry has been desperately short-staffed since forever. As it keeps growing, that shortage has continually been a pain point. This isn't just an annoyance but has caused significant strategic political and economic challenges. Emerging nations like China have realigned their education systems to serve this change well, while North America has continued to lag and suffer. National strategies for global success have to include tech growth and resilience, and that demands a higher level of understanding of tech, and to widen out the talent funnel for these critical positions for both economic growth and digital security. Post-secondary computer science departments haven't been getting the sign-ups that are needed to fulfil industry and government needs for coding talent; improving awareness and interest earlier in education can hopefully help shift the balance to cover this need.

Secondly, a big challenge in tech has been its historical bias. It has had very hard skews toward wealthy white male involvement. Enrolments in undergraduate computer science even in 2016 were 85% male in the United States according to the National Science Foundation. We need to more than triple women's enrolment in CS for parity. African Americans constituted 13.31% of the US population in 2016 (according to Statista) but only 9% of its CS enrolments, meaning, a 50% increase is needed for parity. Computers have been rare, expensive, and obtuse tools, and the historic access to the technology has created significant skews in its use, adoption, and perception. America, and other countries, cannot meet their current or future needs without broadening the base of interest and involvement in computer science. If women and minorities are not actively engaged and encouraged, we will be stuck with what's been dubbed the "leaky pipeline" for talent. This skew isn't simply a matter of shoring up numbers. With technology becoming a fundamental part of everyone's lives, having technology creation so demographically skewed creates additional biases and inefficiencies. The market cannot adequately be served with a lack of authentic understanding through participation. A tech industry without black voices cannot effectively or adequately serve the black public. We've seen artificial intelligence (AI) systems becoming widespread, but we still see skin tone recognition problems in an enormous amount of the products being launched. Right or left political spectrum this is wrong; it's a market inefficiency and a social injustice.

Third, as I mentioned in the first book, the reality is, tech is what will drive the world going forward. Computers and AI are the dominant and revolutionary technology of the 21st century. We need every member of our society to have at least some fundamental understanding of this technological underpinning of our civilization. We need informed and engaged citizens so they can democratically engage in policy decisions with knowledge and consideration. Ensuring basic education provides this understanding of the fundamentals of computer science serves to defend democracy. We need more and better tech-aware policy being made and implemented, and only a greater understanding of the subject can provide a democratic process for those changes to ensure the legal and political system stays in line with the technological developments in society. Public schools can ensure both understanding and access are available for all citizens.

4

Defining Intermediate Scratch

Intermediate Scratch is, in my opinion, the Goldilocks zone of coding education. Not too unfamiliar to be comfortable or too simple to be interesting like beginner Scratch, but not too complicated or boundary-pushing like advanced Scratch can be. Here we have a wonderful middle ground of both capability and comfort. This isn't to say the others can't be, but this level of learning is perhaps the easiest to work with, with less direct step-by-step guidance of beginner Scratch and less difficulty of troubleshooting than advanced Scratch. A rich, broad ground full of potential.

To get to this stage, the student will have achieved the goals of beginner Scratch. This will provide them with a basic toolkit and familiarity with which to explore. We would expect an intermediate Scratch student to understand the fundamentals of movement, know the basic sprite properties, use basic loops and basic conditionals, and have used a few variables. They don't have to have mastery of these skills but should be familiar enough to be able to use them and explore them without much guidance. If you yourself haven't gotten their yet, you may wish to read through *The Teacher's Guide to Scratch – Beginner: Professional Development for Coding Education*, our guide to beginner Scratch.

Intermediate Scratch Attitudes

This is the time when students grow from their newfound comfort of conquering beginner Scratch to exploring it and becoming skilled and independent enough to move into the area of self-guided mastery in advanced Scratch. A wonderful time of seeing students blossom into their own independent creativity and discovery. The attitude I expect students to have is, "I think I can build it". They have the comfort with Scratch to try new things, push themselves, and discover. They won't always succeed without setbacks, second attempts, or assistance, but they'll be able to try a lot on their own, even if their methods might be a little lacking or not have the clarity and simplicity that a mastery of understanding can provide. At this level, they should be failing forward with exploration. With their comfort in Scratch, they can take more leaps and, given time, learn to fly on their own.

The focus at this learning level is exploring – they've seen the parts; now they can see what they can do. Projects will tend to either be large but simple or complex but small. They may take simple systems they've learned and run with them to more fully flesh out their imagination. Or they may take the time to master a particularly complex task in isolation. I often equate coding techniques with tools: the more tools in their tool belt, the more situations they can build solutions to. We want students in this level of learning to explore widely, find new techniques, add them to their repertoire, and test them in different situations. This will often involve a lot of code borrowing, which is often a red flag to teachers used to worrying about plagiarism, but here it's a constructive way to understand new view points and design paradigms. Borrowing code or remixing projects allows us to think outside our own mind, to see others' perspectives and techniques that can be very valuable to understanding the depth and breadth of options that coding can allow. Also, to adequately remix or borrow code, we need to come to understand it and how to adapt it. This provides us wonderful opportunities to both grow and test comprehension.

In general, intermediate Scratch won't deal with complex math or data manipulation. We'll see some math functions, but not deeply complex ones, or at least not complex ones in interactions. Projects will generally still avoid iterative state tracking; projects will have simple states of levels or story points, but not deeply complex interdependent states, like open-world games with dynamic narrative stories and plots. We'll see some interdependence and interactions, but generally not excessive. Programs may have interactions and states but won't have much in the way of special case handling. As a period of exploration, we could see just about anything being explored, but not to the depth and comprehensive interactive understanding we get to in advanced Scratch, where we start looking more toward professional approaches.

Intermediate Scratch Goals

At this level, the goals for intermediate students are the functions that will really allow for larger projects and state tracking. We're setting the stage for complex data handling and interactions in advanced Scratch. We'll know students are ready for the next level when they have a firm grasp on the following goals:

State Machines. The student can have different states in their game, such as menu or Game Over screens and can have scene or level switching affect gameplay.

Clones. The student can use cloning to dynamically spawn objects as needed, with a clear understanding of the role of the original vs. clones, understanding of initiating clones and deleting them.

Logic Operators. Conditionals are being used with •Not, •Or, or •And operators to create more complex conditions. Nested •[If/Else] are used to correctly sort conditions.

Messages. Projects use •[Broadcast [x]] and •[When I Receive (x)] to trigger custom events in code. The student has learned to avoid race conditions by using conditionals or waits.

Reliability. Projects rarely fail due to errors and missteps in code or user input – not that these don't occur, but they remember to test things and find their bugs before someone else does. Though they may need help to fix or avoid them.

5

Intermediate Project 1: Pen Tool Fun

What This Project Is

This project helps students explore the world of geometry and symmetry. Using one of the Scratch extensions, the ✏ **Pen**, we'll start drawing simple shapes and then get a little more sophisticated with nested loops to create interesting patterns. The ✏ **[Stamp]** function will come in handy to make some snowflakes, further exploring the concept of symmetry. In about 45 minutes (for adults), we should end up with an interesting program that reveals some of the beauty of mathematics and the world of science.

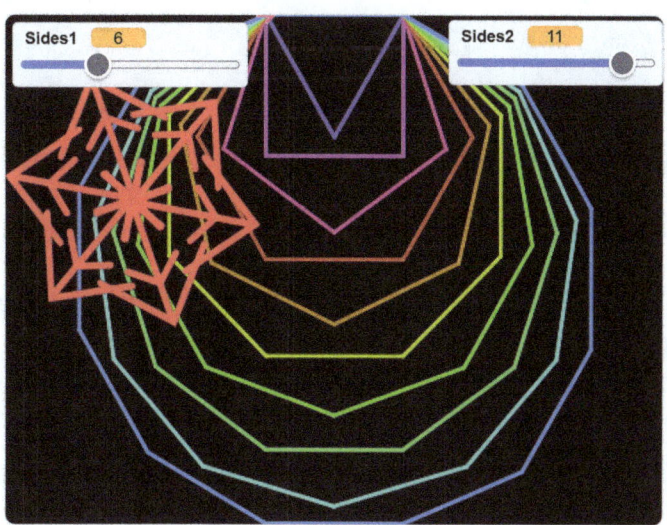

What We're Learning with It

This project shows off one of the Scratch extensions, the ✏ **Pen** tool, the core of Scratch's predecessor – called Logo. This will allow us to draw in-game. •**Motion** blocks are used to control our drawing. By using math, we can explore drawing basic shapes, helping reveal some of the laws of geometry and bringing them to life with real-world application for students. We'll also work with nesting code to build up more complex patterns. Lastly, •**Variable** sliders are introduced as a means of user input.

> **Extensions.** Adding extensions to Scratch, a powerful feature to expand its abilities.
> **Pen Behaviour.** We'll learn the basics of working with the ✏ **Pen** extension for drawing in-game.
> **Precision Motion.** Precision motion commands will be used to make the drawings.
> **Nested Loops.** Nest code tasks inside each other to build up complex pattern behaviours.
> **Geometry Integration.** We'll create a great example of how coding can integrate subject matter, such as geometry.
> **User Variable Controls.** Work with •**Variables**, including allowing users to set and constrict their values.
> **Grid Walking.** We'll learn how to create grids and work through their positions using an algorithm.

Building It

Step 0: Create Your New Project
Make sure you're logged in to Scratch, then click **Create** to begin a new project! Since we won't be using it, we can delete the Scratch Cat sprite by clicking on the trash bin on that sprite's thumbnail in the Sprite Listing.

Step 1: Adding the Pen Extension
Adding extensions is an easily overlooked feature in Scratch. In the bottom left corner of the screen, there's a blue square with a white icon that looks like two code blocks and a plus sign. Click on it to **add an extension**.

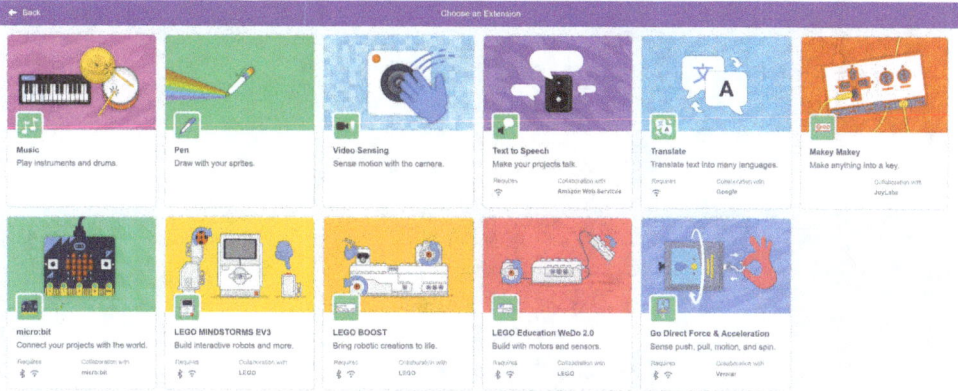

> **Extensions**
>
> *Extensions are a powerful tool for educators, adding enhanced abilities to Scratch. Currently, there are extensions for music (adding more composing options), video sensing (allowing users to work with their webcams), text-to-speech (to vocalize text), translate (to translate text), as well as the Pen tool. In addition, there are a number of extensions allowing Scratch to work with robotics and microcomputer systems commonly used in coding education. These can be great for exploring hardware, robotics, and electronics concepts. In addition, the Lifelong Kindergarten Group at the MIT Media Lab is always working on more functionality for Scratch, so who knows when the next extension may get added!*

We'll add the ① **Pen** extension by clicking on it. This will take you back to the editor with a new category of code blocks now added to your project.

Step 2: Our First Drawing

For starters, we'll need a sprite to use the Pen. We can grab the *Arrow1* sprite from the **Sprite Library**. Note all the **Pen** code blocks that become available when a sprite is selected. After adding it, let's adjust its **Size** in the **Properties panel**, located between the **Stage Window** and the Sprite List. Size is a percentage, so let's type in 25 to make our *Arrow1* only a quarter of its normal size, so it doesn't cover up as much of the screen, so we'll be able to better see the drawings.

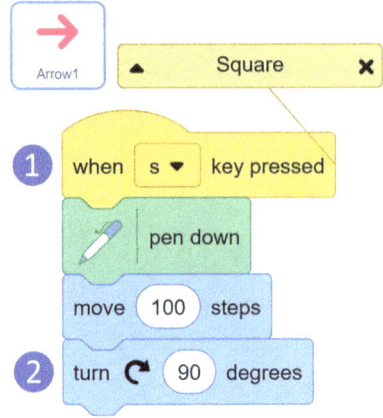

So how do we use the Pen extension? Imagine the Stage Window is a piece of paper. If you put the ✎[Pen Down] and then •[Move (#) Steps], that sprite will draw a line. If you lift the ✎[Pen Up], it'll stop drawing. So we need to switch between ✎[Pen Down] when we want to draw, and ✎[Pen Up] when we don't want to, and then we just use movement to draw the lines we desire. We can switch the **size** and **colour** just like switching what pen/marker we use to draw.

To get things drawing, we'll need an •Event. Let's add a ❶ •[When [Space Key] Pressed] and change it to the "S" key. You'll see why in a bit. Have the ✎[Pen Down] code block below it, followed by a •[Move (10) Steps]. Try pressing the S key a few times. You'll see a thin dark-blue line being drawn behind our arrow whenever it moves. We're drawing!

Let's make things a little more interesting. Increase our •[Move (#) Steps] to 100 and add a ❷ •[Turn Right (90) Degrees] after it. Try pressing the S key a few times. You'll end up drawing a square! But wait – how do you erase things?

Step 3: Reset

Our next job is adding an erase function. This will let us get rid of old drawings (or mistakes). Add a ❶ •[When [Space Key] Pressed]. You could have any other key, but previous projects proved the Space key is handy for resetting things. Under it, add the ✎**Pen** code block ✎[Erase All]. Hit Space and you should get a clean slate. While in the ✎**Pen** code blocks, let's also grab a ✎[Set Pen Size To (#)] code block. By switching the value to 3, our line will be a little thicker, making it easier to see.

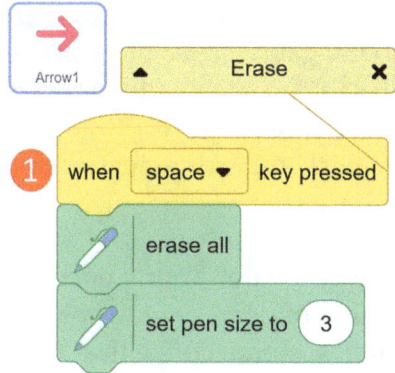

Step 4: Our First Shape

Our "S" key press event is drawing squares. But we want to automate our shape drawing so we don't have to press the button four times every time we want a square drawn. By grabbing a ① •[Repeat (#)] block from •Control, we can set it to 4 repetitions and place our •[Move (100) Steps] and •[Turn Right (90) Degrees] inside it. Now you'll be able to draw a square with a single push of the S key. You can experiment dragging your arrow around to draw more squares around the screen. But if we want our users to be able to do that as well, we need to add a couple of code blocks to allow that. As a designer, we have total control, but users won't be able to move objects around the same way unless we use code to give them that power.

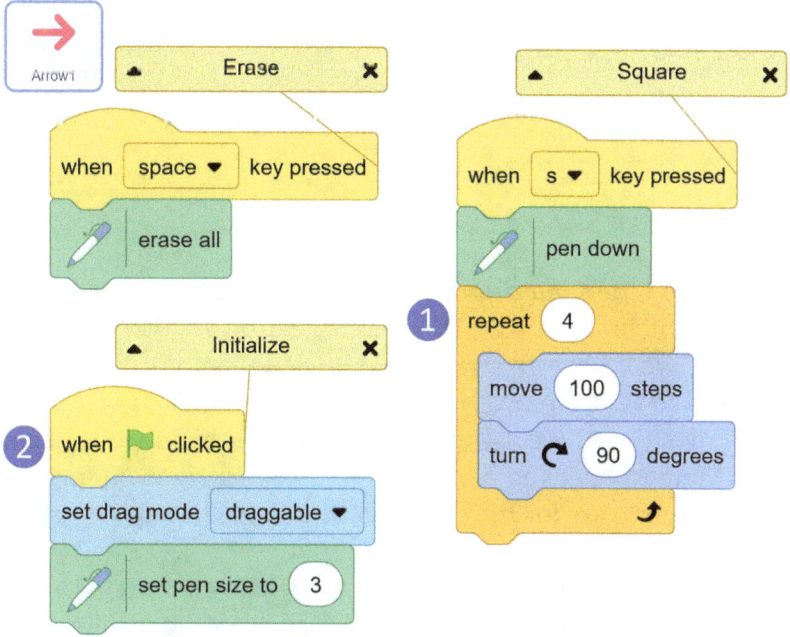

Add a •[When ▷ Clicked] event. Go to the •Sensing category and grab a ② •[Set Drag Mode [Draggable]] code block to connect under it. With this property, the user will be able to click and drag the object where they want in–game. Let's move our ✎[Set Pen Size to (3)] to that event, too, so our users will always get the thicker line drawings even before they reset.

Step 5: Our Second Shape

See how easy it is to draw a square? Your students will probably be able to give you the correct number of degrees to turn in order to make a square if you ask them when building a project like this. But what about a hexagon? Let's try adding a hexagon drawing routine to our project. We can simplify life by simply duplicating our •[When [S Key] Pressed] event. We covered three methods for duplicating code in *The Teacher's Guide to Scratch – Beginner: Professional Development for Coding Education*; hopefully by now you have a favourite! I like right-click, duplicate for same object copying.

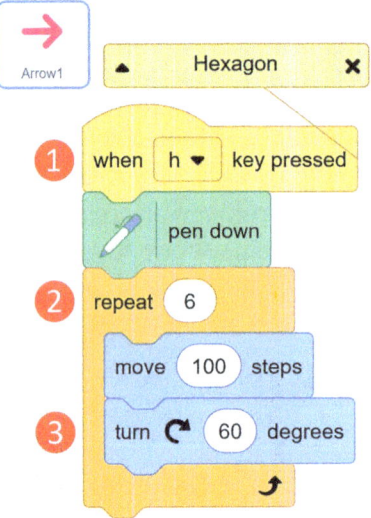

Now that we've got our duplicate, we can start editing it to make a hexagon. Start by switching the ① •Event to trigger on an "H" key press. Then, for a hexagon, we'll have to switch our ② •[Repeat (#)] to 6 for six sides. We'll also need to switch our ③ •[Turn (#) Degrees]. Can you guess the correct number of degrees? We use the inside angle for this, so for a hexagon, the •[Turn (#) Degrees] needs to be 60 degrees. Shapes give you a good opportunity to quiz your class for the right side and degree numbers or have them calculate them.

Note that when we draw our shapes, the •[Move (#) Steps] is the length of an edge, not the diameter of our shape. The more sides to our shape, the bigger it gets, since we're using the same •[Move (#) Steps] value.

Step 6: A Complication

At this point, we should be able to tweak the number of repeats and turn degrees to make any basic shape, except one. How do you make a circle? A •[Repeat (#)] with 0, 1, or infinity repetitions doesn't give us great results. In Scratch, we'll have to fake it. Since we're dealing with limited Stage Window resolution, we can simply create an object with enough sides to create the illusion of a true circle. Start by duplicating either shape script.

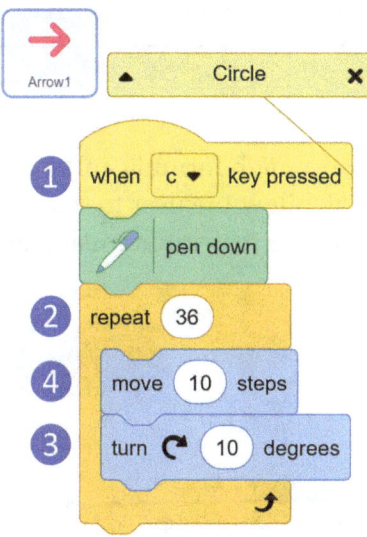

First, change the ① •Event to a "C" key press. Change our ② •[Repeat (#)] to 36, as this is more than enough sides for what we can fit on screen in Scratch, and it makes the math easy. Change the ③ •[Turn (#) Degrees] degrees to 10: 36 × 10 will give us the (360) degrees to complete a circle. If you test this, though, our •[Move (100) Steps] is WAY too much to fit on-screen. By cutting it down to ④ 10, we'll easily stay on the screen and complete a circle. You could fit up to a 30-step circle on-screen, but you would start noticing the corners at that scale. Also worth noticing, if you play around with equivalent multiplications, you'll see that the •[Repeat (#)] also has an aspect of time, so the larger the repeat, the longer it takes to draw.

> **Over the Edge**
>
> *You'll notice that drawings get messed up if they hit the edge of the Stage Window. Scratch has a number of protection protocols that sometimes mess things up for us, but they're there to help. Objects are prevented from moving off-screen so that you don't lose track of them. They will, under most conditions, stay in a place that you'll still be able to click and drag them around. This helps preventing young and*

new users from losing track of objects and getting frustrated. By preventing things from moving off-screen or shrinking too small, Scratch tries to be more accessible and understandable to young and new users. If objects could just move off-screen as soon as people connected a ●**[Move (10) Steps]** and ●**[Forever]** code block, we'd have objects marching off to infinity! So protections like this one try to make it easier to work with, but that does mean we have to deal with things possibly getting messed up. As long as we rescale our drawings or reposition our sprite before it starts drawing, we should be able to avoid any problems. It might seem frustrating, but what would the point of drawing off-screen even be? You'd never see it!

Step 7: Variable Sides

For our next step, we'll use a math formula and ●Variable to make our shape method universal. This will allow the user to select what shape they want to draw and dynamically adapt to their choice using geometric laws.

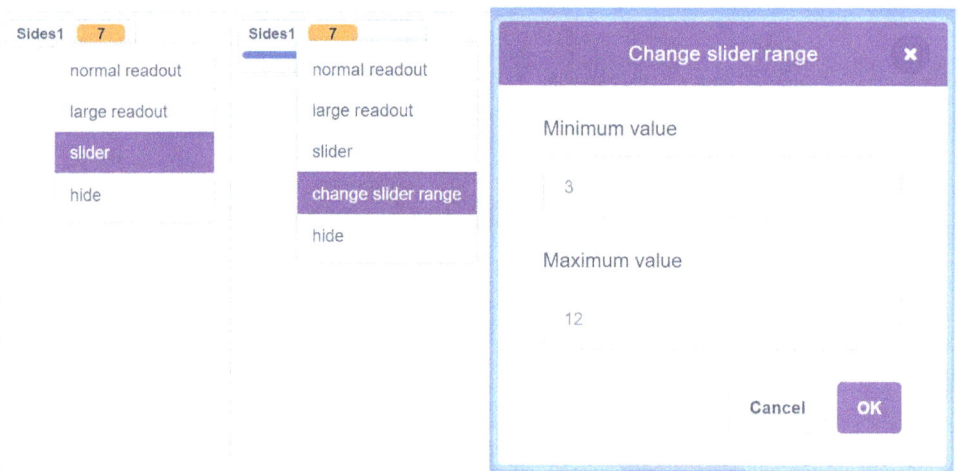

Start by duplicating one of our other shapes code blocks and setting its ❶ key press to "X". Then go to the ●Variables category and *make a variable*. Call it ●*"Sides1"*. We'll keep this ●**Variable** visible but change how it's displayed in the game. Go over to the ●*"Sides1"* variable display in the *Stage Window* and right-click on it. Choose "**Slider**". It now displays as a slider control that allows the user to select a number between 0 and 100. But wait! We know a 0–2 sided shape isn't going to work, and we know any large number just ends up looking like a circle anyways. So let's right-click on it again and select the option **"Change Slider Scale"**. That allows us to set the minimum value to 3, and the maximum value to 12.

Intermediate Project 1: Pen Tool Fun ◆ 19

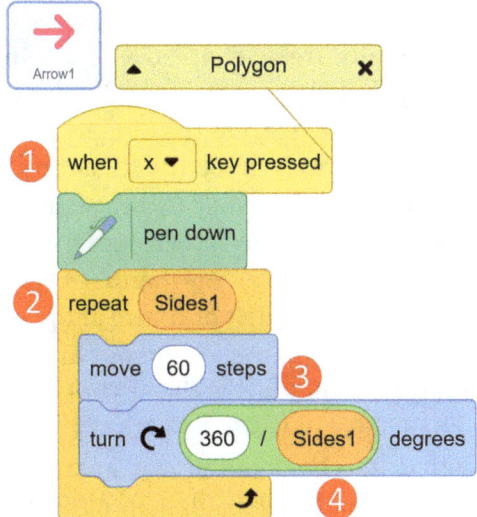

Now we'll take our ② •("*Sides1*") code block and place it in the •[**Repeat (#)**] as its value. This allows our user to determine how many sides the shape should have. However, we're going to have to change the •[**Turn (#) Degrees**] degrees to match this selection. We'll need the •((0)/(0)) (Division) •**Operator** code block. If we place our ③ •((0)/(0)) (division) code block into our •[**Turn Right (#) Degrees**] code block, we can type in (360) in the first value and place our second ④ •("*Sides1*") code block in the second value. Our •[**Turn (•(360)/• ("Sides1")) Degrees**] combo now dynamically adapts to the number of sides.

> **Geometry**
> *While we always use the phrase "there's 360 degrees in a circle", it's actually true of any closed shape. All the interior turns will always total 360. You have to get back to where you started to close a shape. So when we're making regular polygons (equal-sided shapes), we can use the formula 360/Sides to determine the angle of the turn at each of the corners.*

Try experimenting with •("*Sides1*") values using the slider and then pressing X to see them drawn. The •[**Move (100) Steps**] can be too big for anything above 12 sides (if you position the arrow at Y: 180, the top of the screen just fits), but we'll deal with that a little later. Maybe adjust it to •[**Move (60) Steps**] for a happy middle ground. Also note that if you select the slider, you need to click the background to have your key presses trigger events again instead of being intercepted by the slider control.

Step 8: Shapes Sequence

Now that we can dynamically create shapes, let's use this ability to create a sequence. Start by copying the //Polygon stack, and change it to a ❶ •[When [0] Key Pressed] event. In this sequence, we're going to draw standard polygons in order from triangles up to a dodecagon (or 12-sided shape).

We're going to need all the space we can get, so let's start by placing our arrow in a specific spot to know how much space we have. We'll add the •Motion code block ❷ •[Go To X:(#) Y:(#)] and set the coordinates to X:50, Y:180. This will place Arrow1 a little left of centre at the very top of the screen. Grab a •[Point in Direction (90)] to make sure our arrow is facing right. Both these code blocks should be at the start of the •Event. But running the •Event, you'll see one of the challenges of working with the ✎Pen extension. We've just drawn an unwanted line by moving our arrow! To avoid that, add a ❸ ✎[Pen Up] code block as the very first thing in our •Event. This way, we can •[Move (#) Steps] without drawing, get set in the right position, then have the ✎[Pen Down] before drawing.

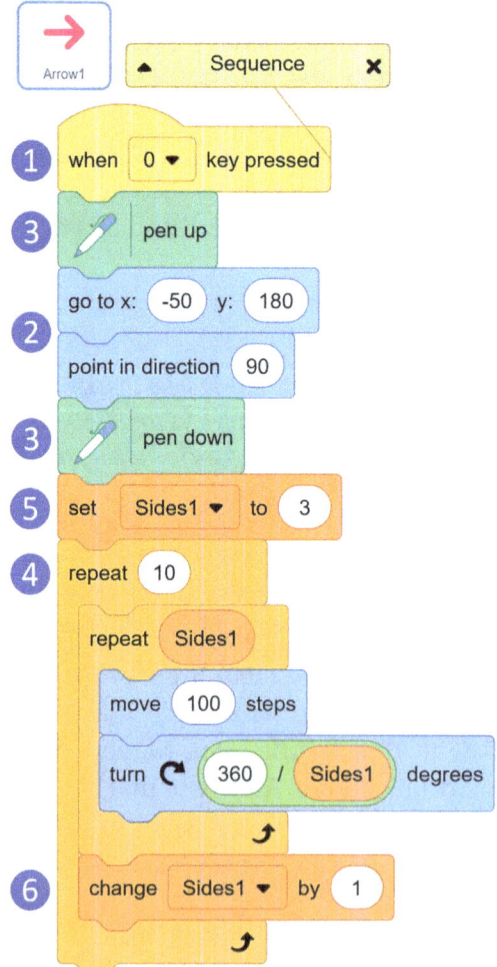

For drawing a sequence of shapes, we'll need to run our shape drawing routine more than once. The ④ •[Repeat (10)] is perfect for it. Place it so the existing •[Repeat •("Sides1")] is inside it. Now we can draw ten shapes in a row, but we still need to change what shape we're drawing.

For this we need to work with our •Variables. We want to start by drawing a triangle, our simplest closed shape. To start the sequence with a triangle, change our •("Sides1") variable by placing a ⑤ •[Set [*variable*] to (0)] code block above the •[Repeat (10)] and selecting the •("Sides1") variable with a set value of 3. This will start our sequence with a triangle. Now, to change it, grab a ⑥ •[Change [*variable*] by (1)] code block and place it inside the •[Repeat (10)] after, but not inside, the shape drawing •[Repeat •("Sides1")]. Switch its variable to •("Sides1"). This will increase the shape's sides after it completes drawing a shape. It will let us cycle up through each shape in order. If we change it to •[Move (100) Steps], it will just fit inside the Stage Window. Try it out!

Step 9: Nested Shape Patterns

We created our first drawing pattern combining ten standard polygons. Let's get creative with shape combining patterns. We'll use the same •[Repeat (#)] inside a •[Repeat (#)] concept, but with set shapes rather than a sequence of shapes.

To get started, duplicate the •[When [0] Key Pressed] stack and change the ① •Event to the "1" key. Now we need to do a little surgery (to change a few things around). Remove and delete the •[Set ("Sides1") to (3)] and •[Change ("Sides1") by (1)] code blocks. Move the ② •("Sides1") *variable* code block from the inner •[Repeat (#)] to the outer (first) •[Repeat (#)], replacing the "10" value. Duplicate (copy) the •[Move (100) Steps] and •[Turn Right (•(360)/•("Sides1")) Degrees] code blocks and place them inside the outer •[Repeat (#)] underneath the inner •[Repeat (#)] (and not inside it).

For this sequence, we're going to use two different shapes. That will require another variable. Head to the •Variables category and **make a variable**. Call it •"Sides2". We need to make sure •"Sides2" is set up like •"Sides1"; right-click on the variable display in the Stage Window for it and select **Slider**. Then right-click on it again and **change slider range** to minimum 3 and maximum 12. You may want to move the •"Sides2" display to another corner of the Stage Window so it doesn't get in the way of the drawings.

Place a ③ •("Sides2") code block as the value for our second, inner •[Repeat (#)] block. You'll also need to swap the •("Sides1") code block in the inner •Repeat's •Turn combo with a •("Sides2") code block. This will let that stack draw the second shape. We can add in a ④ •[Wait (0.1) Seconds] code block in the inner •Repeat if we want to slow it down to watch the drawing

22 ◆ Intermediate Project 1: Pen Tool Fun

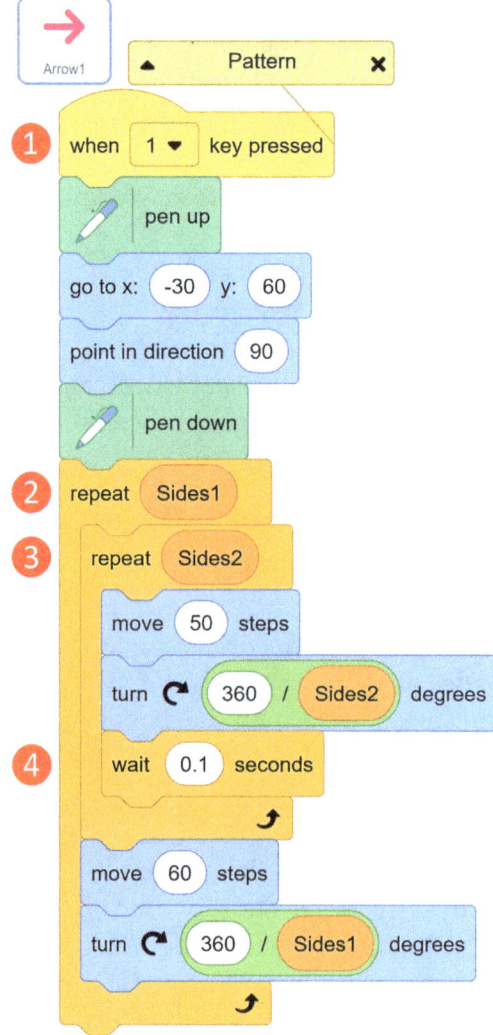

process. Though if *"Sides2"* still equals 0, we won't see much happen, so be sure to set it to another value to test it.

Now there are two last little things we want to change. Since this can result in pretty big drawings, let's switch one or both •[**Move (#) Steps**] code blocks in this stack down to 50 in order to fit the drawings on the screen. You can have the two moves have different values to make duplicate sides give patterns instead of just drawing a single shape over the top of itself. We also want to centre our drawing more, so switch the •[**Go To X:(#) Y:(#)**] code block to X: -30, Y: 60. This should keep our drawings fairly well-centred.

With that, you can now try out your new nested shape pattern drawings. You'll discover some amazingly beautiful patterns, including patterns we see in nature. This can be a great way to explore symmetry, phyllotaxis (leaf growth patterns in plants), and morphogenesis (the shape of biological systems and their growth – especially flower shapes). Be sure to try changing both variables to see how they combine to make different patterns!

Some really amazing things can happen with emergent design and complex patterns. Learning to combine drawing with the clone system can unlock outstanding results. You can get started working with clones in the Butterfly Catcher game in Book 1: Beginner.

Step 10: Randomizing Function

It can be a lot of fun finding interesting patterns, but systematically working through the combinations can take away some of the surprise. Let's add a function to randomize the number of our •*"Size1"* and •*"Size2"* variables so we can generate random patterns with the press of two buttons (one to randomize, the other to call our existing draw function).

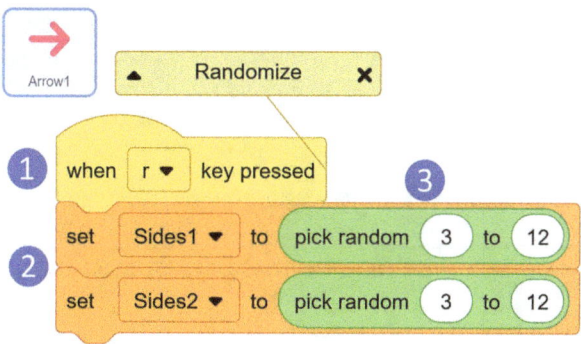

We'll need a ❶ •**[When [*Space*] Key Pressed]** event, and we'll change it to an "R" key press, for "random". Next, we'll need some •Variables code blocks. Get two ❷ •**[Set [*variable*] to (0)]** and put them in this event. Set one to the •*"Sides1"* variable, and the other to the •*"Sides2"* variable. Now we'll need the ❸ •**(Pick Random (#) to (#))** code block from •**Operators**. Put one in each of the •**Set** code blocks. Use the values 3 to 12 in both of them. You could experiment beyond 12, but I find they aren't as interesting as the smaller numbers.

Try randomizing your shapes by pressing "R" and drawing them out with "1".

Step 11: Stamping Patterns

Now that we've explored the line drawing a fair bit, let's turn our attention to another **Pen** function – the **[Stamp]** code block. It allows us to add a new sprite and draw it from scratch. Hover your mouse over the **Choose a Sprite** button in the bottom right-hand corner and you'll see a series of buttons extend out the top of it. Click on the brush icon to create a new sprite with a blank **costume**. You'll be automatically taken to the **Costume** tab to draw it. Name it *"Snowflake"*.

Switch to the **Costumes** tab. We'll need to draw using the **Line tool** to make our snowflake. Drawing a pure white object on a white **canvas** is a little awkward, so I suggest using an off-white/pale-blue colour to make life easier. Click on the **Outline** box and select a **colour** you like (you can always change it after). The number on the right side of the Outline box can be used to select our **outline thickness**. I suggest increasing it to 10. We need to draw just one axis of our *snowflake* because we will be using automated rotation and stamping that will copy it to make the whole.

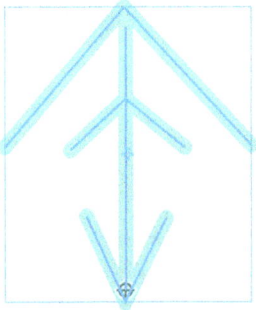

Start by clicking on the centre point of the canvas, then extend the line upward. This is the base axial line of your *snowflake*. You can add some lines extending off this line to the sides to make it more complex. Try a combination of lines flat across, angling upward, or angling downward off the base line to see how they'll look in the final stamped design. You don't need too many and can always come back and adjust things after seeing how it works out. If you've used a bright colour to draw, you can select all your lines now and switch the **colour** to white/off-white now.

Before we're finished with art, let's change our background. White on white or even off-white on white is a little hard to see. Click on the **stage** on the right side of the screen. This time we don't need to **choose a backdrop**; simply click on the word "Stage" or "Backdrops 1". When you select the **stage** instead of a Costumes tab, you'll see a **Backdrops** tab. Go to the Backdrops tab so we can alter the art of the backdrops. All we need here is a single backdrop that's a solid black rectangle larger than the visible area. Set the **fill colour** to black and draw in a big rectangle (using the **Rectangle tool**), covering everything. The whole **Stage Window** area should now be black.

Now that we've got our art, we can add in our code. Switch back to the **Code** tab. Grab a ① •**[When [*Space*] Key Pressed]** event code block and switch it to the "W" key (for "winter"). Under it, use a •**Motion** code block, ② •**[Go To [*Random Position*]]**, so when we press the button, a snowflake will be drawn at a random location in the Stage Window. Next, we will use a pair of ③ •**[Show]** and ③ •**[Hide]** code blocks. We want to •**[Show]** the snowflake to use the ✏**Pen** tool, but we don't want our single-axis sprite appearing outside of that. For now, stack them together and we'll add our draw code in between them.

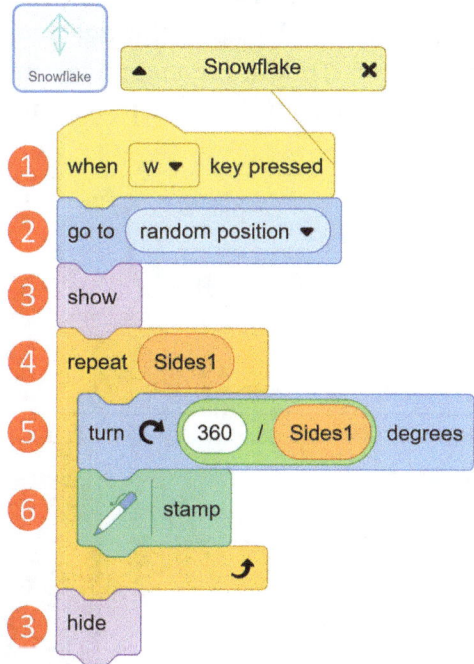

Let's get down to drawing! Remember the same basic principles we used in our shape drawing. A ④ •**[Repeat (#)]** loop will run •**(*"Sides1"*)** times. Our snowflake will be dynamically drawn based on that variable. Inside a •**[Repeat (#)]** (note: make sure the •**[Hide]** is placed under the •**[Repeat (#)]**, not inside it), use a ⑤ •**[Turn Right (#) Degrees]** code block, and inside that place our formula, •**((360)/•("*Sides1*"))**, so get your division •Operator code block and adjust its first value to (360) and put a •**("*Sides1*")** code block in its second value. Because we aren't using line drawing, we now use our new ✏**Pen** code block: ⑥ ✏**[Stamp]**. ✏**[Stamp]** doesn't draw lines; instead, it will draw an exact visual copy of the sprite, calling it at that location. Try it out. You now get a snowflake drawn with a number of points equal to the •("*Sides1*") variable. The [Stamp] block will just make copies of your sprite's current costume in the geometric pattern we've defined through code. Feel free to change up your costume design now that you see it in action.

See how the symmetry uses your costume design to make a more complex pattern. Change the •"*Size1*" variable (or randomize it by hitting "R") to make more or less complex snowflakes. Real snowflakes tend to have six sides, but we don't have to stick to that rule here.

Step 12: Grid Terrain Generation

Now that we've seen how convenient [Stamp] can be to work with, we'll give one more demonstration of its use. In this step we'll use what's called a "grid marching" algorithm that will step us through every tile in a grid. We'll use that to make a mosaic or random tile map; this can be a fun procedural generation method to make random world maps for games, but they also just look nice.

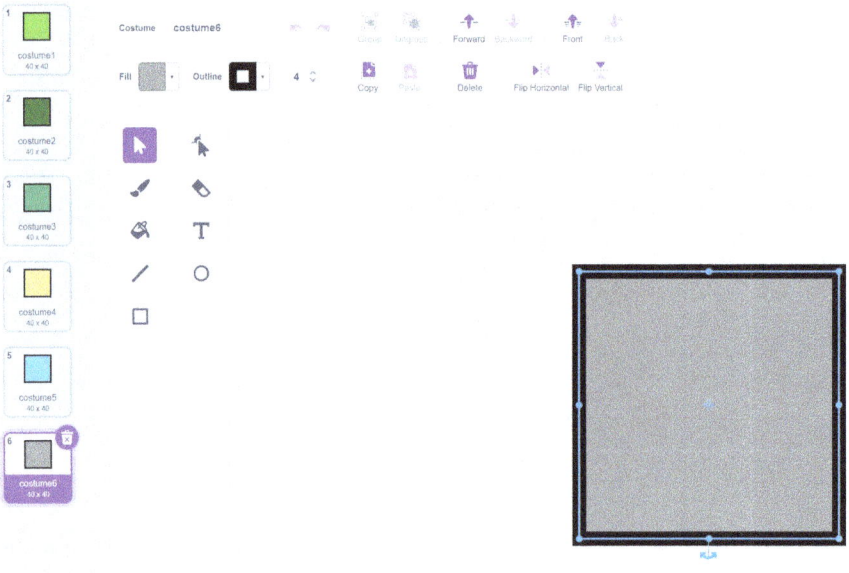

For this we'll need to create a new blank sprite, so hover your mouse over the **Choose a Sprite** button and then click on the paintbrush among the icons that spring out of the top of the button. This new sprite we'll name "*Tiles*". In the **Costumes** tab, we'll make some simple costumes. Just use the **Rectangle tool** to make a square of a solid colour; you can keep the black outline to create a grid of black lines or turn it off by setting it to 0 width or transparency, if you prefer. The square needs to be exactly 40 pixels by 40 pixels. You can see the current size in the **Costumes List** on the left-hand side. Once you've got one made, you can right-click on it in the **Costumes List** and choose to **duplicate** it. We'll make six different costumes, each on a different colour. I went with a natural colour scheme to represent different terrains – light-green grasslands, dark-green forest, dark-turquoise swamps, blue water, yellow desert, and grey mountains – but you can choose your own palette.

With that, switch back to the **Code** tab and we'll add in our functionality. We'll start with a ❶ •[When [G] Key pressed] – G for grid. Now we want to make a copy of our tile stamped across the whole screen, so let's start in the top left corner with a ❷ ✏[Pen Up] and •[Go To X: (-240) Y: (180)]. Then we'll need to •[Show] our tile; we have to show it in order for ❸ ✏[Stamp] to work, but we want the original parent sprite to disappear when it's not stamping, so we'll [Hide] it at the bottom of the stack. Now we get to the heart of the grid marching system. We'll need a ❹ •[Repeat (10)]; this will allow us to stamp out each row of tiles. Our screen is 360 pixels tall, so you'd think we'd only need nine repetitions of our 40 pixel sprite, but we're centring things, so we'll have half tiles all along the edges, which means we need one extra row. Inside we'll need a •[Set X To (-240)]; this will ensure each row begins from the leftmost position. Now, how do we draw each row? Well, it's going to take more than one tile, so we need another ❺ •[Repeat]. This time we'll set its value to 13, 480 pixels wide divided by a 40 pixel tile, plus one extra for the half tiles. For now we'll keep it simple, so inside this repeat, add a ❻ ✏[Stamp] and a •[Change X by (40)]. This will stamp the tile and then move to the right. Done in repetition, it draws the whole row. Now, below the •[Repeat (13)] but inside the •[Repeat (10)], place a ❼ •[Change Y by (-40)]. This will move down to the next row after it completes drawing the row. Then, at the very bottom, add a •[Hide]. Try it out! You should see a 130 tile drawn in a perfect grid. Except it's all the same tile . . .

So that's the basic system for stamping out a grid; you could replace the ✏[Stamp] with any kind of function you needed to do in each cell, and you could resize things to different numbers of rows and columns by adjusting the numbers. It's a very important and powerful technique. But for today, we want to make a map. So let's add a bit more code. Inside the •[Repeat (13)], above the ✏[Stamp], let's add yet another ❶ •[Repeat]. This one we'll set its value to a •(Pick Random (1) to (6)), and inside it we'll place a •[Next Costume]. This way, on every tile, it will randomly pick a costume. Try it out. Now it's starting to look like a map!

Of course, as we're just generating things at random, we get a very random, or noisy, map. Let's add one more little bit of code to smooth things out. There are lots of ways one can choose more natural results, but here we'll just try a very simple little tweak, which also happens to be a handy little technique to know about for lots of different uses. We'll need an ❷ •[If <Condition> Then] code block. For its conditions, we'll put together two code blocks

Intermediate Project 1: Pen Tool Fun

Tiles — Generate Map

1. when `g` key pressed
2. pen up
   go to x: -240 y: 180
3. show
4. repeat 10
   set x to -240
5. repeat 13
   - (2) if pick random 1 to 2 = 1 then
     - (1) repeat pick random 1 to 6
       - next costume
   - stamp
6. change x by 40
7. change y by -40
3. hide

you might not expect. First, we'll get an •<(0) = (0)> and we'll change its second number to (1). In the first number, we'll add a •(Pick Random (1) to (2)) code block. This basically acts like a coin toss, running whatever is inside the •[If] half of the time. Of course, you can change the numbers to get different probabilities, or change from an = to a < or > to test different ranges. The important thing is, we've created a random chance of things happening or not, and we didn't even need a variable! Now, we'll take this •[If] and we'll place it inside the •[Repeat (13)] and stick the •[Repeat] with the •[Next Costume] code block inside it but the [Stamp] below it. Now it will change what costume it stamps only half the time; this will make more streaks of colour/terrain rather than be totally random. Check it out! Your own random little worlds!

If you enjoyed exploring graphics like this, you might want to try expanding purely abstract art into representative art. In Book 3: Advanced, we explore data visualization in the Bar Charts and Data Files project.

Step 13: Customization

A very fun feature we haven't touched yet is pen colour. Now that we've got that dramatic black background, let's play around a little more with pen customization.

First, check out the [Set Pen (Colour) to (Colour)] code block and add it to any of your *Arrow1*'s square, circle, and hex stacks. You can make each one a distinct colour. The ① [Set Pen (*Colour*) to (*Colour*)] code block above the •[Repeat (#)] will let you set a colour for the shape. You could also set each one's [Set Pen Size To (#)] value to something different to see how those values affect things.

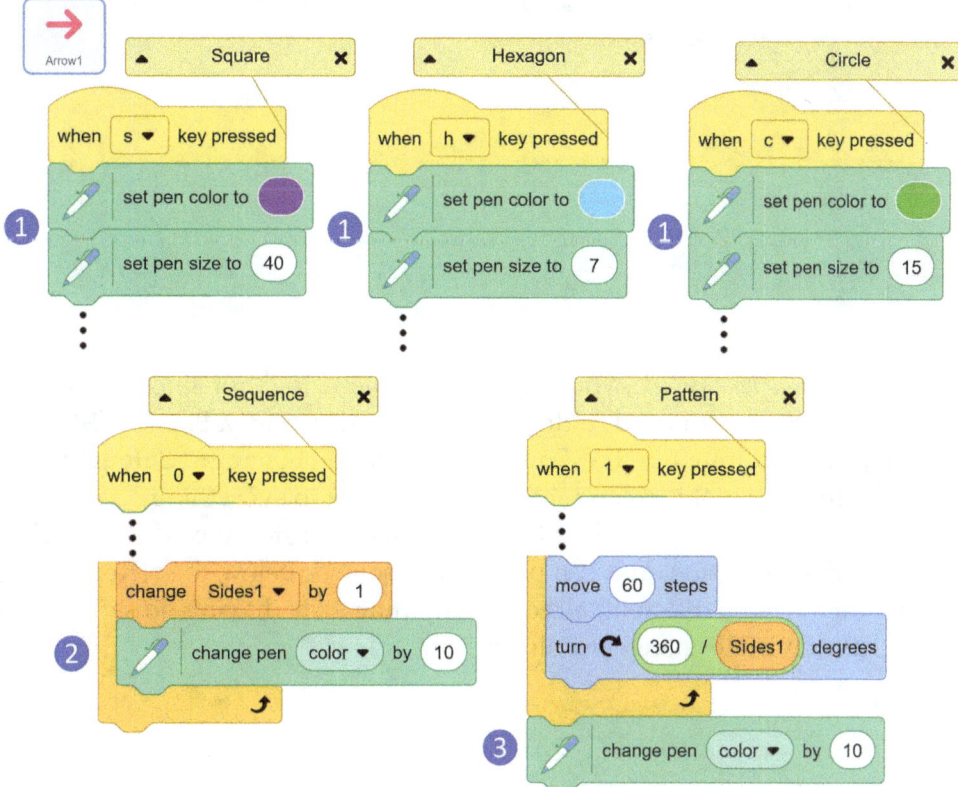

Next, let's get some more colourful patterns by grabbing a ② [Change Pen (*Colour*) By (10)] code block and adding it to our •[When [*0*] Key Pressed] just above the •[Change (*"Sides1"*) by (1)] code block. This will ensure each shape is a different colour; it really helps distinguish each shape! We can also throw a ③ [Change Pen (*Colour*) By (10)] at the end of the •[When [*1*] Key Pressed] events so that each time it runs, it'll draw a new colour. In the •[When [*Right Arrow*] Key Pressed] event, you could try using two [Set Pen (*Colour*) to (*Colour*)] code blocks to make the inner and outer shapes distinct colours. There are lots of options to explore.

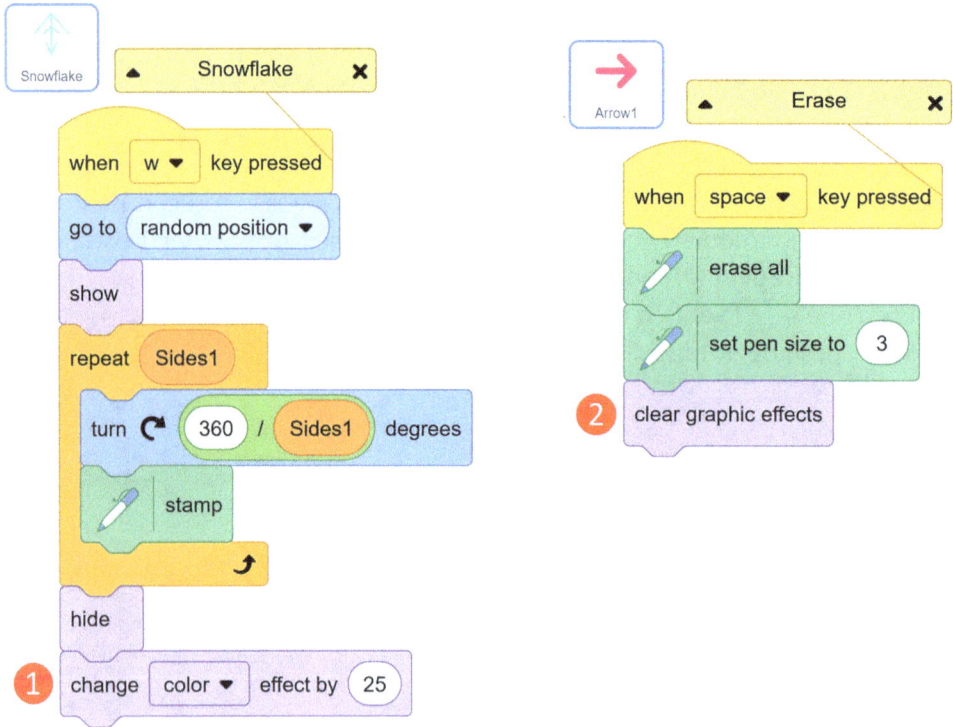

In our *snowflake* or *tile*, you can explore colour too, but since it uses the [Stamp] function, you need to do things a little differently. Being a [Stamp], to change their colour, use the •Look code block ❶ •[Change (Colour) Effect By (25)]. To remove this colour shift from applying to everything else after it, you'll need to have a ❷ •[Clear Graphic Effects] to stop the effect, possibly in your //Erase function. You can place that above or below the •[Repeat (#)] so that each will be different, although anything grey (having no saturation) won't have a visible change in colour!

With that, we'll end our project exploring the ✎ **Pen** extension. There are lots of wonderful creations to explore with these methods, so feel free to experiment and extend it!

For working copies of this and every project in the book series, visit www.massivelearning.net *for direct links to Scratch projects, and to see our other projects and resources for coding education!*

6

Intermediate Project 2: Interactive Story

What This Project Is

Our next intermediate project is an interactive story. Here we'll make a picture book–style story, but we'll be using animations to bring it to life. Importantly, we won't just make a single story; we're going to include user input so they can drive the path the protagonist takes through our fantasy story. Here we'll be engaged in telling the story of a wizard caught without his magic wand and having to wander through a fantasy world without the use of his magic. Can he make it safely back to the castle? The user will have to make the right decisions, with multiple possible endings to the story. In a little over one hour (for adults), we'll build out an interactive story with multiple scenes and characters, user input guiding the flow of the story, and multiple animations bringing our story to life.

What We're Learning with It

This project will really expand on the scale of the projects we're doing. Here we're not just using Scratch to tell an animated story; we're telling an animated story with multiple paths and endings. We'll reuse the same fundamental techniques to build out the sections of the story, but within different sections, we'll take the opportunity to learn some additional techniques. We'll learn about the concept of switches or keys through the use of a simple

Intermediate Project 2: Interactive Story ◆ 33

inventory. We'll practice our animation techniques, including text displays. We'll explore using size scaling in our animations to create a perspective effect. We'll work with an array of backdrops and deal with scene, or room, switching. This project provides a great base for any interactive story, allowing you to use this project as a training module that can allow your students to then create their own interactive stories.

> **Text Input.** We'll use the •[Ask] code block to get text answers from our user, opening up a huge new field of user interaction.
> **Saving Answers.** We'll learn to save the user's answers and how to handle text variables.
> **Multiple Choice.** We'll create multiple-choice questions, but we'll also learn how to handle multiple answers with nested •[If] statements.
> **Narrative Paths.** We'll learn to think about narrative paths and how to diverge and converge story paths.
> **Messages.** We'll use the message event system, opening up a new world of possible event triggers.
> **Inventory/Switches.** We'll have the player gain items and learn ways to track and react to these conditions.
> **Fade Transitions.** We'll explore some animation techniques to fade to black and fade from black to give a cinematic experience to our story.
> **Layering.** We'll learn about layering in both the costumes and in code to ensure our assets look right and our graphical effects work right.

Building It

Step 0: Create Your New Project
Make sure you're logged in to Scratch, then click **Create** to begin a new project! Since we won't be using it, we can delete the Scratch Cat sprite by clicking on the trash bin on that sprite's thumbnail in the **Sprite Listing**.

Step 1: Goblin Storyteller
Now that we know the basics of working with Scratch, we're going to get a lot more brief with our instructions. We're assuming you've learned where things are, and you can recognize the categories of code blocks, so we're going to get a lot more shorthand with instructions, so we can pack in more code and more interesting projects and focus on techniques and methods rather than the logistics of getting code blocks.

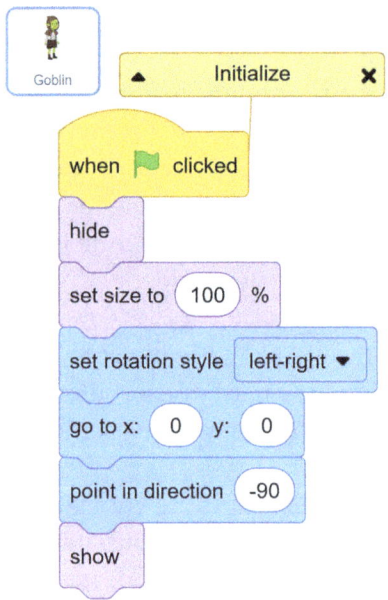

We'll set our background to the **Forest** and add the *goblin* Sprite. This character will introduce the story to the player. Let's start by setting things up, or initialize them. This sets things up to start the game correctly, similar to the Reset function we used in *The Teacher's Guide to Scratch – Beginner* projects. This will simply run under the •[When ▷ Clicked] event. We'll •[Hide] our sprite, then arrange it correctly, then set it back to visible with •[Show]. The •[Set Rotation Style [*style*]] code block, here set to the **"Left-Right"** setting, is great for having characters switch to face left or right, but not rotate to any direction like the default setting. This way, characters can walk around, left or right, without standing on their heads. It isn't actually necessary to

Intermediate Project 2: Interactive Story ◆ 35

hide, set, then show the character, but I think it's a good habit to enforce for future coding skills.

Step 2: Messaging Events
With our *goblin* in place, we're going to learn the fundamental technique to our interactive adventure – the use of *message* Events. Messages, or broadcasts, are Events that you trigger in code, rather than from a specific external stimulus. Instead of waiting for a user input, these trigger when you tell them to. This will allow us to tell the computer when to switch scenes, run an animation, or do whatever, exactly when we want it to. This lets us break our different scenes into their own events and determine in code and gameplay when they should happen.

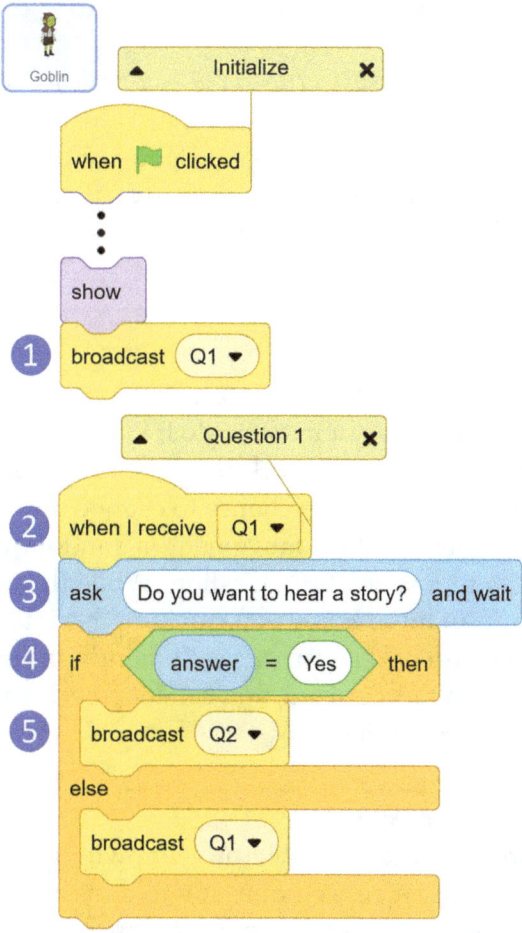

Go to the Events category. The ① **[Broadcast [*message*]]** block is used to initiate an Event. Then any object, sprite or stage, can have the listener ② **[When I Receive [*message*]** event. This will trigger when anything

Broadcasts that message. In the dropdown on either block, you can either choose from the existing messages or choose to create a new one. Here we'll use *"Q1"* and *"Q2"* for the first and second question the *goblin* is going to ask.

With the messaging blocks added, we can go to Sensing and add the new and powerful ❸ [**Ask (*question*) and wait**] code block. This code block combines a text display; enter the text you want to show, and it then allows the user to type in an answer and hit Enter to continue the program. This code block breaks the sequence, waiting until the user finishes typing (by hitting enter) before it will move on. Whatever the user types becomes a built-in variable, called *"Answer"*. The ❹ (**Answer**) block is available in Sensing, and we can test it just like we would a normal variable. So here we confirm that the player wants to hear a story.

If they answer "Yes", ❺ the game will move on to **"Q2"**, the second question from the *goblin*.

Messages are an amazing system to know how to work with. If you want to learn more, the Batty Flaps project in Book 1, The Teacher's Guide to Scratch – Beginner, *had a great system to allow a game to restart without pressing the* ▷. *In Book 3,* The Teacher's Guide to Scratch – Advanced, *we'll show you how messages can be used to create a procedural generation system that will create endless gameplay!*

Step 3: Wandering

Of course, it's pretty silly to start an interactive story by asking if the user wants to hear a story. There's not much to do if they don't, ❶ but we wanted this as an example of redirecting back to a question. Let's have our *goblin* wander around until the user has a chance to change their mind.

We add in ❷ some random movement, a random delay, loop it in a repeat, and then ❸ repose the first question. While this might seem a little unnecessary, it's important to know you can jump back to events; they don't have to be in linear order. This can be really handy in quizzes or guessing games. If an answer isn't recognized, you can loop back and ask the question again – a very handy technique!

Step 4: Handling Human Answers

So now we're going to address the big problem with human input: it's very human. Even if you pose a yes/no question to a human, how often do you hear "yes" or "no" as the answer? In conversation, not that much; you could get "yeah", "uh huh", "sure", "nah", "maybe", etc. Language is complex; people are unpredictable. This poses a huge challenge when we have to simplify and predict everything into our code. Here we show one of the ways

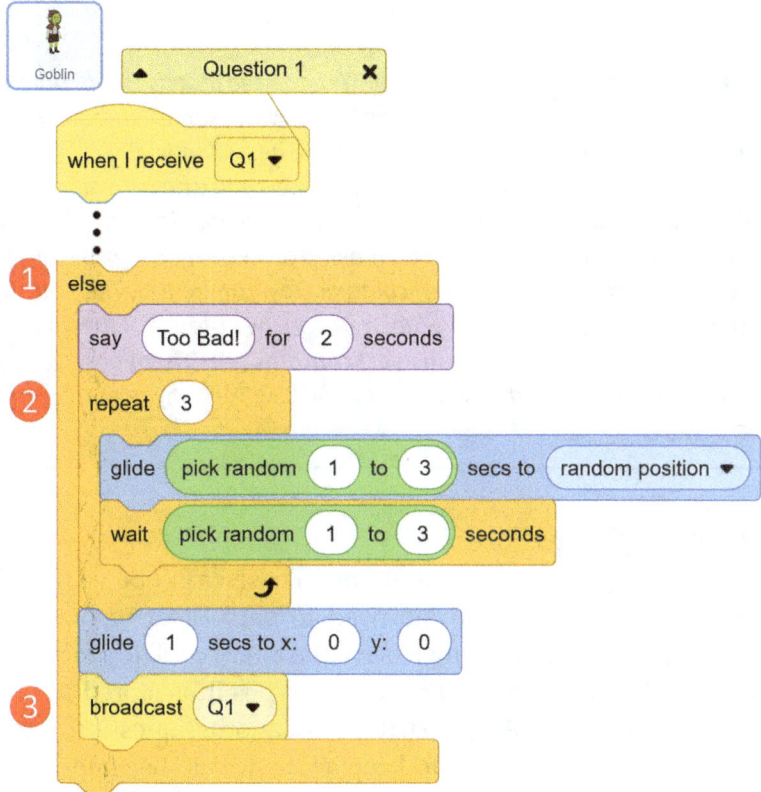

of handling things. Using the ●<<> or <>> block and stacking them inside each other (it's tricky – it'll take some practice), we can account for multiple answers. Here we've given a number of popular answers, but even this isn't exhaustive. This is why having the aforementioned technique, or reposing a question, can be really important. If you don't know how to handle the input from the user, maybe just try re-asking. Ideally, we'll handle all the likely answers, but this is a good backup. We can also try another technique to improve things, as we'll see in a few steps.

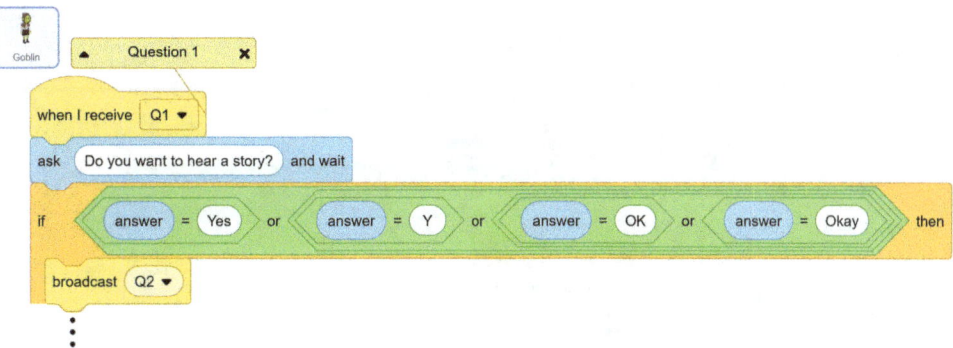

Testing Text

Handling text in programming can be tricky. You can see that the list of potential answers can be huge, and this is just for one language! In some computer programming languages, capital and small-case letters will be distinctly different, so "Yes", "yes", and "YES" would all be different distinct answers you'd have to account for! Thankfully, Scratch makes our lives somewhat easier by not distinguishing between capital and small-case letters. You can try making answers more flexible by testing if answers include text using the <(text) contains (textfragment)>. This code block will search the text given to it, such as (answer), for any instance of the text fragment you provide. So you could try ●[If <(answer) contains (y)> Then] and "yes", "yeah", "y", "yup" would all be true, but so would "nay", "yellow", or "why does this not work right?" So beware!

Step 5: Personalizing Things

We'll add a second and third question now. These won't be very picky. We'll take whatever the user types in; no need to parse it or respond here. The important thing is that we're going to save the answers. The problem with the ●[Ask] code block is that whatever the user types in is stored in ●(Answer), but since there's only one ●(Answer), if we ask another question, it overwrites the previous answer. If you want to keep an answer to use later, you need to put it into a standard ●Variable. Using ① ●[Set [*variable*] to ●(Answer)] will allow you to save the user's answer to a specific ●Variable that will let you call on it later, which is what we'll need with these two answers. Now, we also want to hide the variable displays so they aren't cluttering up the game view. In the ●Variables category, you'll see checkboxes next to the new variables. If you uncheck those, the variables won't be displayed in the game, but we can still use them in the code. We will want to hide (uncheck) all the variables we use in this project so they don't distract the player.

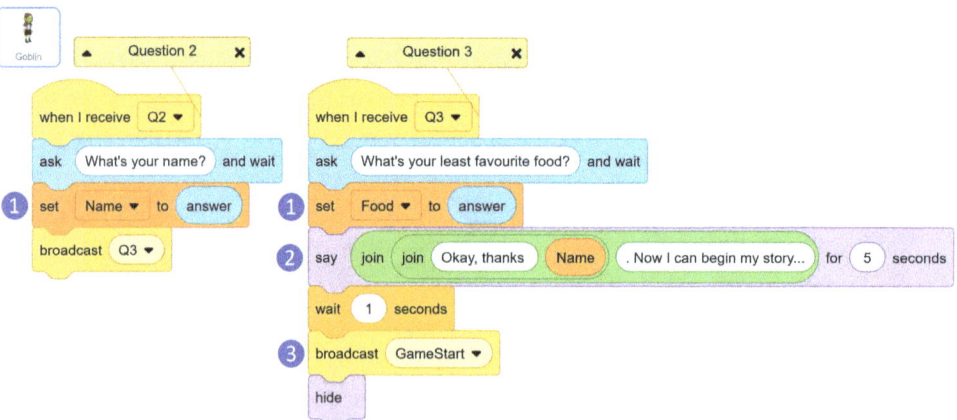

After our third question, we'll actually use one of our saved answers. Now that we know the user's name, we can call on it to use it in our program. Here we use the ② •(Join (text1)(text2)) code blocks to put together a three-part sentence, joining text with the user's name, and then joining that with a final comment. These answers will let us personalize our interactive story. Last, we'll ③ •[Broadcast ["*GameStart*"]] as a message event to begin the story and •[Hide].

Step 6: Our First Scene Objects

We can now begin work on our first story scene. We'll set up a *griffin* sprite and a *wizard* sprite from the library. We need to establish some ① initial code to position and size them for their scene, but importantly, we need to •[Hide] them. Everything but our *goblin* starts invisible and only becomes visible for its specific scenes. In the *griffin* we'll add a new message event, ② •[When I Receive ["*Transition*"]], which will act as a catch-all event that will turn everything invisible again, so when we change scenes, everything disappears and only the objects for that scene will be made visible. Our *wizard* doesn't need the "*transition*" event since as our player character protagonist, they'll always be a part of the scene.

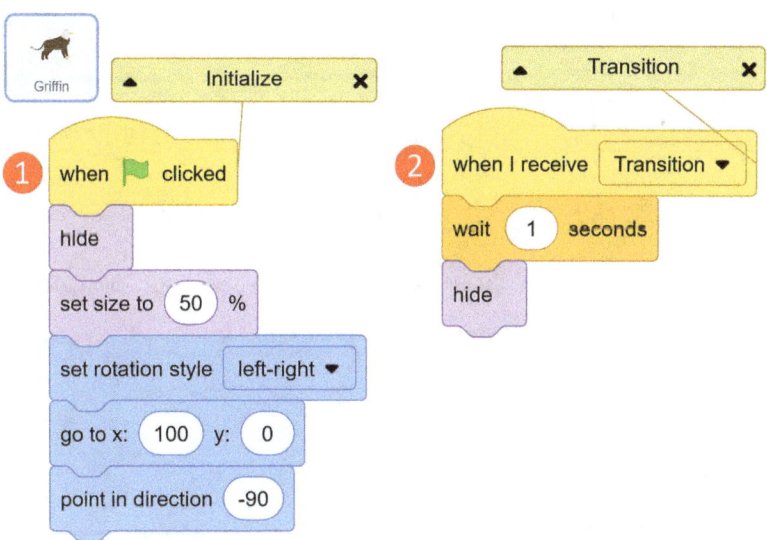

While we're setting things up, let's switch to the stage and add in the different scenes we'll need. We can add the six **backdrops** we'll need for the story: **Mountain, Desert, Farm, Jungle, Castle 1,** and **Castle 3**. We can also set up the messages that will call each. You may notice that we could have used the •[When Backdrop Switches To [*backdrop*]] event when the scene

Intermediate Project 2: Interactive Story

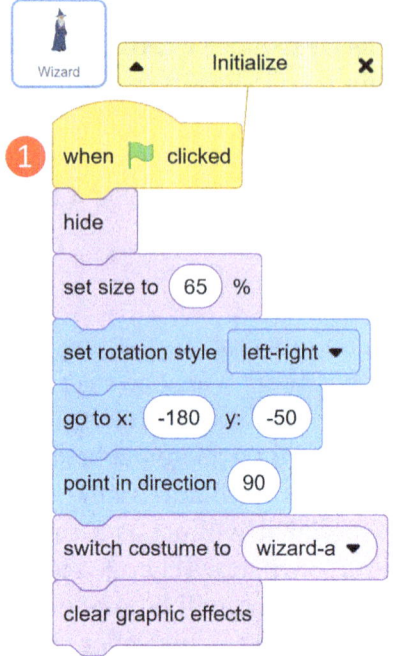

has different backgrounds, but if some scenes reuse the same **backdrop**, you need to use separate events for each. Since messages can be independent of the **backdrop** changes, they allow us to have multiple scenes with the same **backdrop** if needed.

Step 7: The Wizard Sets the Story

With the *wizard* added, we can now begin the actual story. We'll add a new message event, •**[When I Receive [*"GameStart"*]]**, that triggers the first scene after the *goblin* is done with their questions. Here we'll have the *wizard* animate by switching costumes and provide the player with a monologue to set the story in motion. We'll use our •**(Join)** technique to again place the user-entered name into the dialogue.

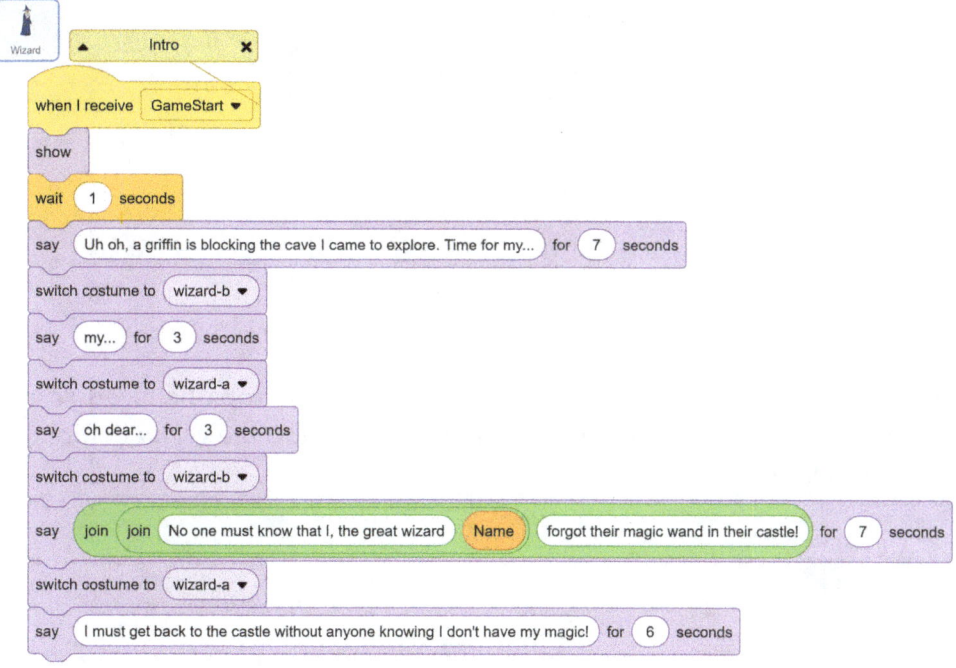

We can add in some animation to our *griffin* as well to help it set the mood, showing itself active and intimidating, posing a problem, and setting the story in motion.

At this point, we can add a sprite to control our cinematic transitions. Hover your mouse over the **Choose a Sprite** button and select the **Brush** option above to create a new sprite without a premade costume. We'll name this new sprite "*Fader*". In the **Costumes** tab, we just want a basic black rectangle larger than the viewable screen. We'll use this to fade to black and back between scenes. You will need to set its position to X: 0 and Y: 0. The ❶ *"Transition"* event will make it run a repeat sequence to change its ghost effect to ❷ fade in then ❸ fade out. We want to make sure that we have a ❹ •**[Go To [*Front*] Layer]** code block to make sure this sprite will be drawn overtop (in front of) all other sprites in the game. Now, our transitions will

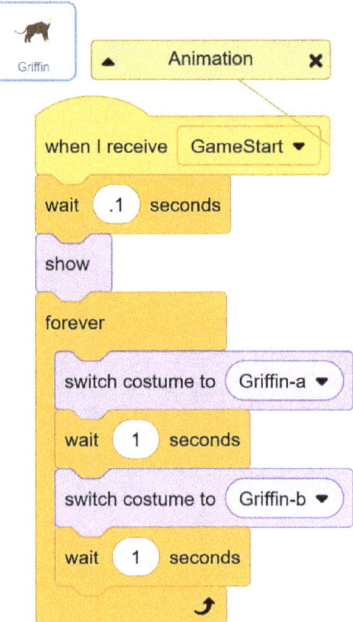

beautifully hide all the switching objects and backdrops between scenes. You can add a ❶ [**Broadcast** [*"Transition"*]] event to the *goblin* before calling *"GameStart"*.

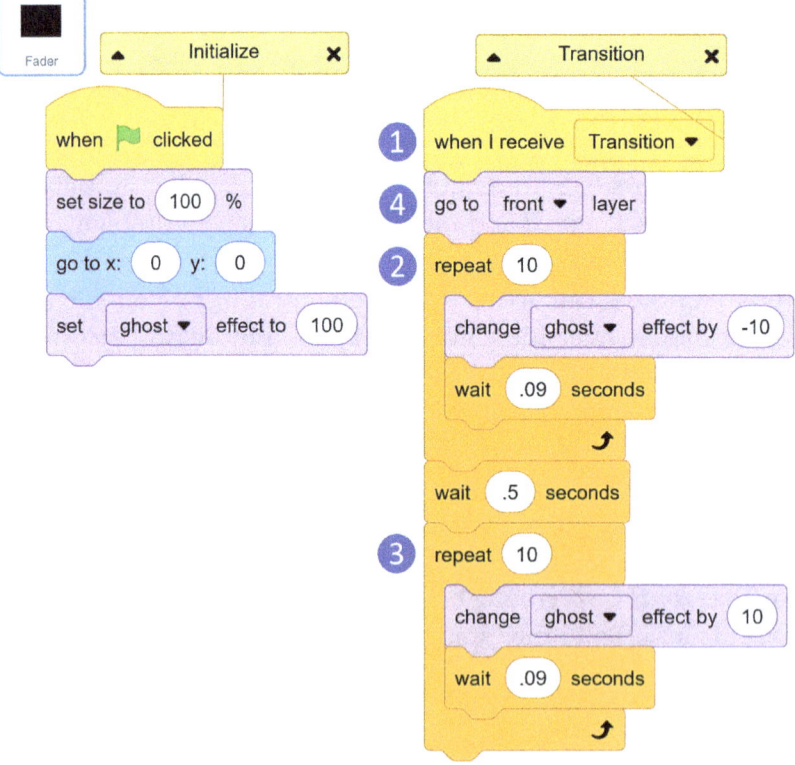

Intermediate Project 2: Interactive Story ◆ 43

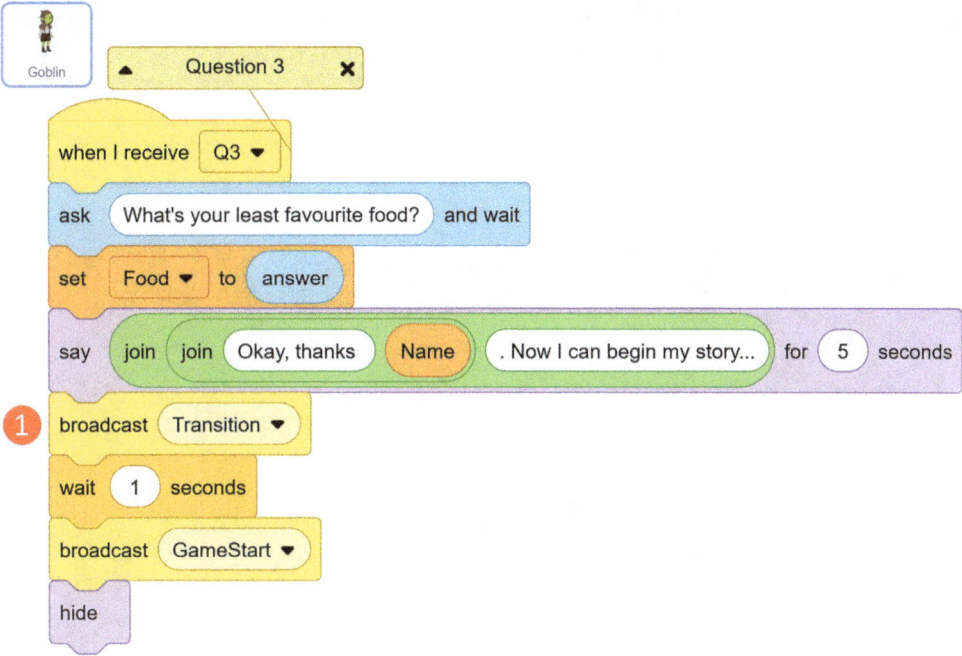

Step 8: A Fork in the Path

With the story established, we come to our first player input in our narrative. The *wizard* needs to get back to the castle; now we'll put a choice of options in front of the player and use an ●[Ask] to get their answer. We'll do this with a *"PathChoice"* event, so we can refer back to it without our intro.

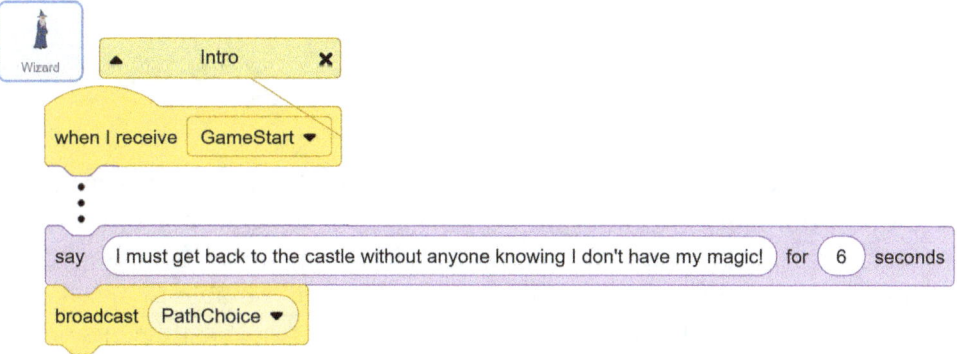

Our ❶ ●[Ask] here uses the extra technique I mentioned earlier. Here we provide the user with some guidance about what input we want from them. Instead of having the player enter a word or phrase to describe what they want, we reduce the choice to a number of set values. This helps us reduce our work; we can avoid needing exhaustive lists of options and still include our exception handling recursive function to return to the question if they don't give us an expected answer, but by guiding the user, in this case with letter prompts, we can simplify the process for both of us.

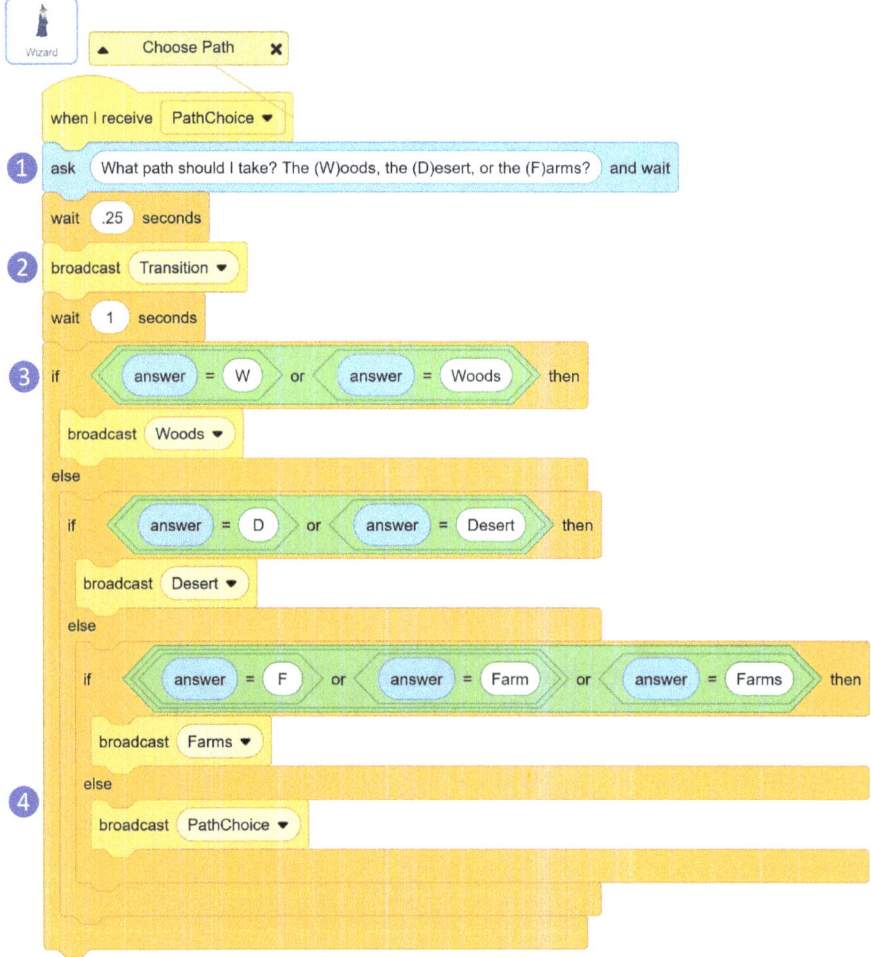

Getting the answer, we set the ❷ **"transition"** event in motion (creating a fade-out), and we check the •**(Answer)**. If it matches one of our three expected answers, we send out a message calling that scene; if it fails all those options, we send them back to the start so they will be able to answer the question again. You'll notice that here we have our ❸ •**If** statements nested inside each other. Sometimes it's important to have your Ifs separate, one below the other, but in this case, we need to make sure they nest one inside the other. The reason is, we need the final ❹ •**Else** to have excluded all the other answers. If each •**If** is separate, the •**Else** can run even if one of the other expected answers is true. By nesting our •**If**s, we only end up at the final •**Else** if all the answers have failed. This difference is really important to

Intermediate Project 2: Interactive Story ◆ 45

understand. We'll use nested and unnested •**If** chains in our projects; knowing why and when to use each is really important if you want to be able to get the right answer. We could have used unnested •**Ifs** if instead of •**Else** we made our "re-ask the question clause" a separate •**If** and used the •**Not** and •**And** logic •**Operators** to piece together the condition of the answers •<•<**Not** •<•(**Answer**) = (**W**)> and •<**Not** •<•(**Answer**) = (**Woods**)>>>, etc.

Step 9: Off to the Farm

With our first choice made, we now have three separate options to follow up on. We'll start by making our Farm scene. We need a ① •[**When I Receive** ["*Farms*"]] event to begin the scene. The *wizard's* code has a large ② •[**Wait (#) Seconds**] block in it because they will be conversing with another character and need to •[**Wait (#) Seconds**] for that other character to finish their dialogue to respond to it. We can also add in a second ③ •[**When I Receive**

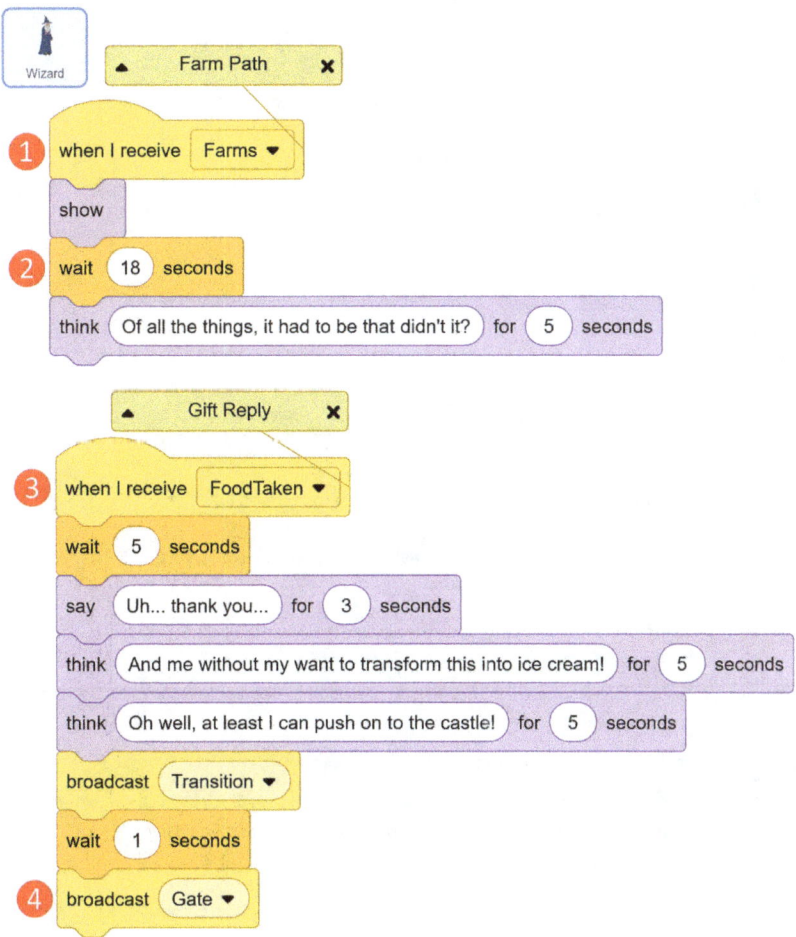

["*Message*"]], this time for a *"FoodTaken"* message. This will be a second event within the same scene. After some dialogue, we'll start the transition and ④ call the next scene. There's no multiple choice here; this is what a straight narrative path looks like.

To make sense of the *Wizard's* dialogue in this scene, we'll need to add our *elf* sprite. We need our basic start and transition code that we can copy from the *griffin* and tweak. Then we need the scene message. ① **[When I Receive ["*Farms*"]]** will ensure that the *elf* responds to the scene change and becomes visible and active just when we need them. Here we have the *elf* talk with the *wizard* and offer them some food! Sadly for the user, it's exactly that dish they hate. Oh well, a gift is a gift. We'll add in the ② *"FoodTaken"* event to add in a little more dialogue.

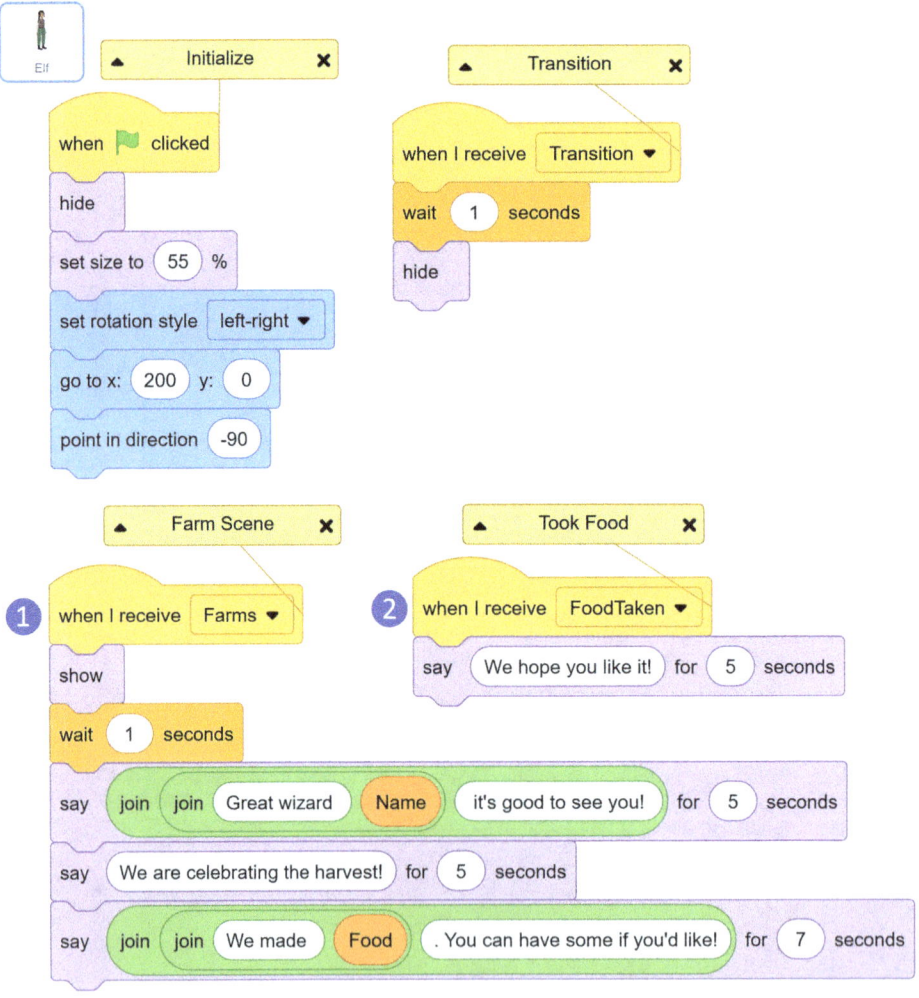

Intermediate Project 2: Interactive Story ◆ 47

Lastly, we need to add in a *gift* sprite so the *wizard* can receive the food from the *elf*. This gives a good example of an inventory item, acting as what's known as a switch or key. This is an item the player can acquire that will change the game later on. Our *gift* needs the usual •[**When** ▷ **Clicked**] event setup, but we're adding an extra •**Variable** here to track that the player has or doesn't have the food: •*"HasFood"*. We ① set it to 0 to begin the game to indicate they don't have it.

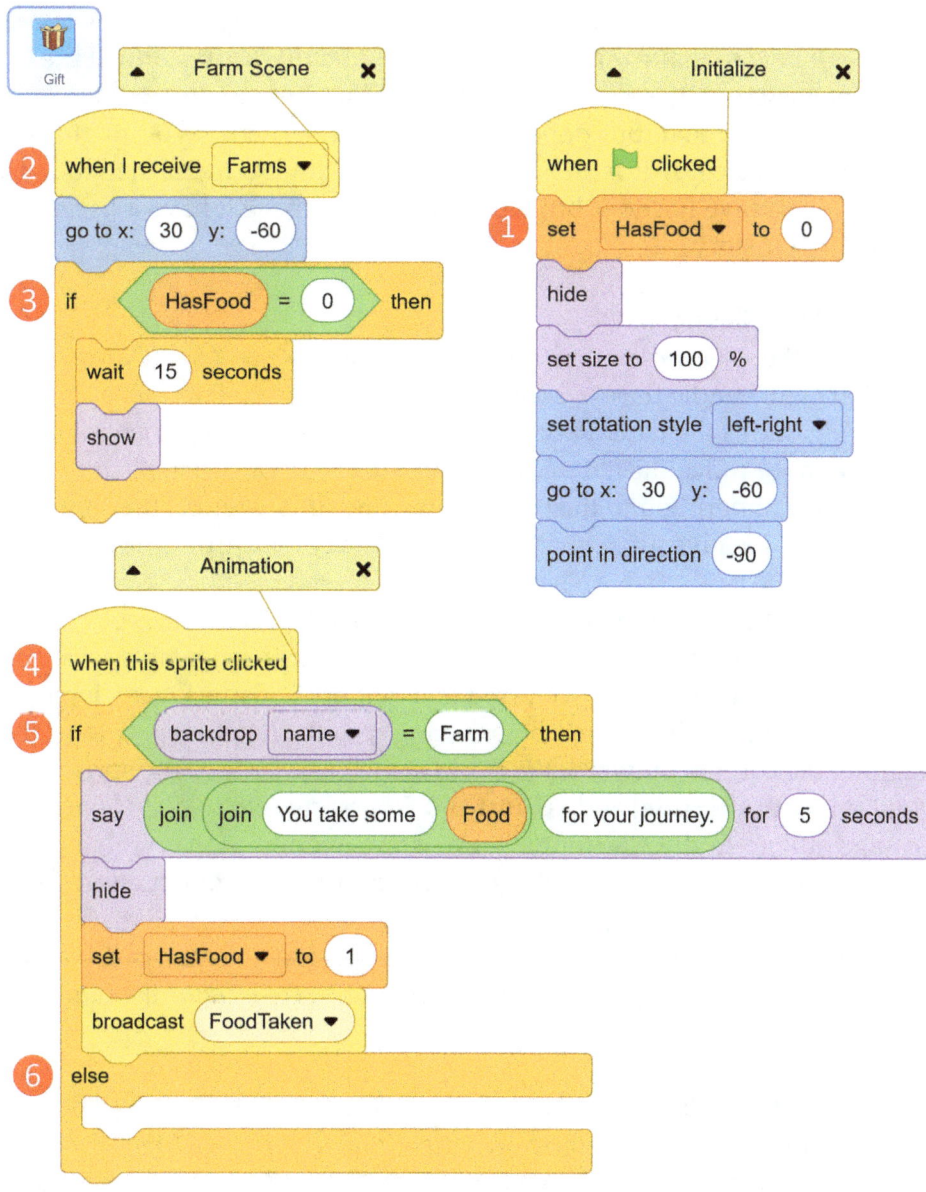

The ②・[When I Receive ["*Farms*"]] event makes sure the *gift* only shows up in the right scene, but we've added an extra ③ ・If here. The ・If checks that the player doesn't have food to show. This isn't necessary in this situation, since the player can't end up back here in our story, but in a story where the player could return to a place, it's very important to have conditionals like this so you don't have nonsensical duplicate objects; it only generates if it hasn't been taken yet. The "*GiveFood*" event shows the *gift* when the *elf* calls for it.

We also add in an interactive element here of the *gift* being ④ clickable. We could have had the transaction be automatic, but again, here's a good example of how to make things interactive. ⑤ It will display only if the current scene matches the scene where it's found (in this case, we use the **backdrop** name to test that, but depending on the game/story, you might need a different method). If it is the initial scene, we'll give them the food. The ⑥ else, we'll leave for now, but we want it because we'll have a use for the *gift* later in our story.

Step 10: Into the Desert

We'll leave the ・"*Farm*" path for now and switch to the ① ・"*Desert*" path. The *wizard* will handle the ・"*Desert*" event all on its own. This scene provides a loss endgame as the *wizard* becomes lost in the desert. It uses an animation sequence to ② move the *wizard*, ③ changing their scale to make them look like they're wandering into the background, and after a while, they ④ begin to fade away. We give a ⑤ final message. ・[Hide] the sprite so we can reset the graphic effects.

To handle the Game Over, we'll add a new blank sprite – *GameOver*. In the **Costume** tab, we'll make a solid black rectangle over the viewable area. Then we'll use the Text tool to add some red text that says "Game Over", large and centred. You can scale the text by selecting it with the **Selection tool**, then dragging the corners outward. We'll add some basic code to the sprite. We'll ① start it with a ・[Hide] and ・Ghost 100. If it receives the ② ・"*GameOver*" message, it'll go to the ③ front layer to draw overtop of everything else in the game; it'll ・[Show] and then ④ slowly decrease the **ghost** effect. When it's done, it will ⑤ ・[Stop All], ending the game.

If you like animation but are looking for an easier way to get started, our Dino Dance Party in Book 1: Beginner offered a great easy start to working with animation and sound.

Step 11: Into the Woods

The last of the three initial choices was the ・"*Woods*". In the ・"*Woods*" scene, the *wizard* exchanges some dialogue with a *witch* character and then proceeds

Intermediate Project 2: Interactive Story ◆ 49

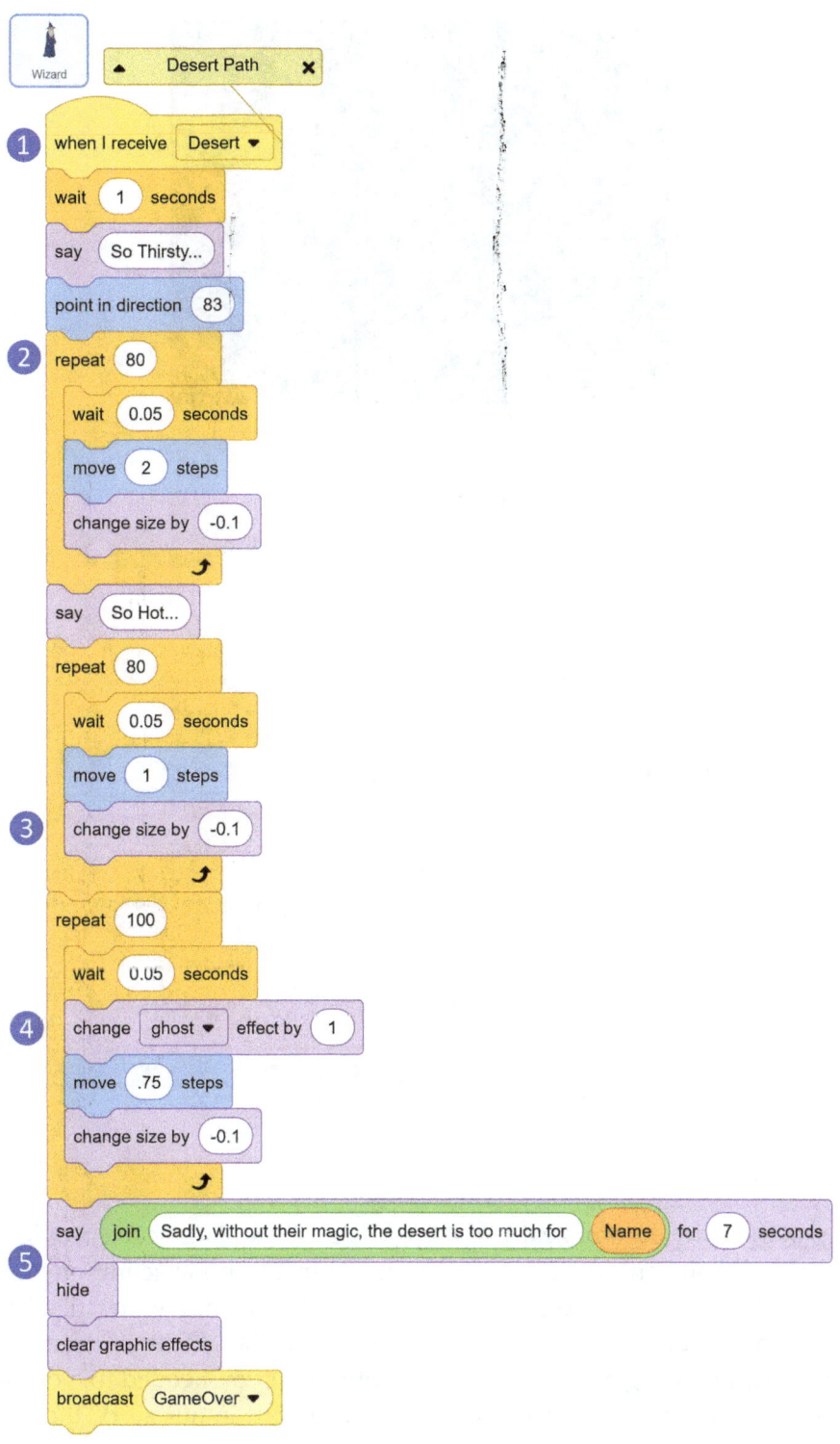

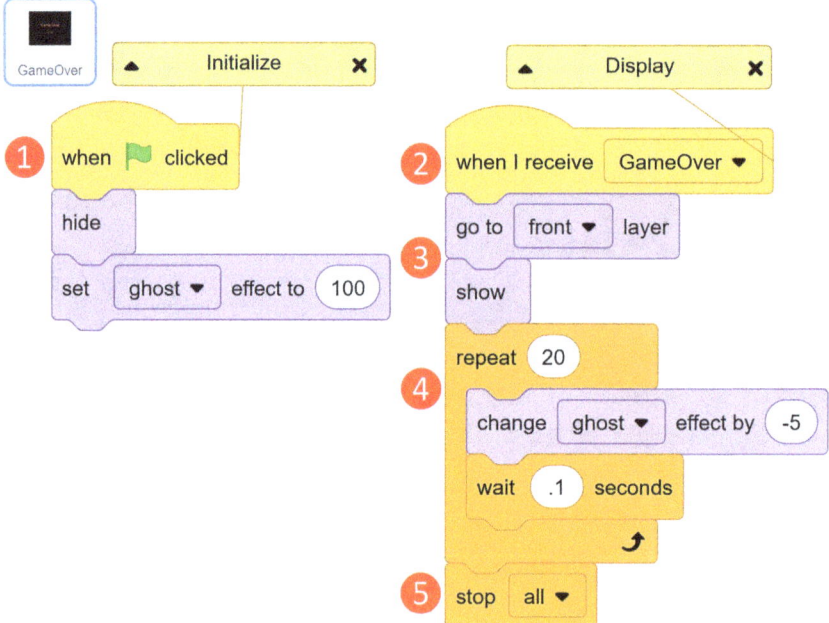

along in the story automatically to the ●*"Gate"* scene. This scene provides an example of an inventory change without a direct interface component. The *wizard* will be given some potions by the *witch*, but they won't need to be clicked on, nor are they shown, unlike the *gift* in the ●*"Farms"* scene. Here we actually are using them as the default option. If you needed to, you could create a ●*"HasPotions"* variable and have it set, but in our story, the *Wizard* must either have the food or the potions, so we only need to track one of them and we can assume the other.

Intermediate Project 2: Interactive Story ♦ 51

We can add the *witch* sprite to our game to interact with the *wizard* here. The *witch* needs the same basic beginning as the *griffin*. We'll add a little bit of dialogue to mention the potions the *wizard* has gained in this encounter.

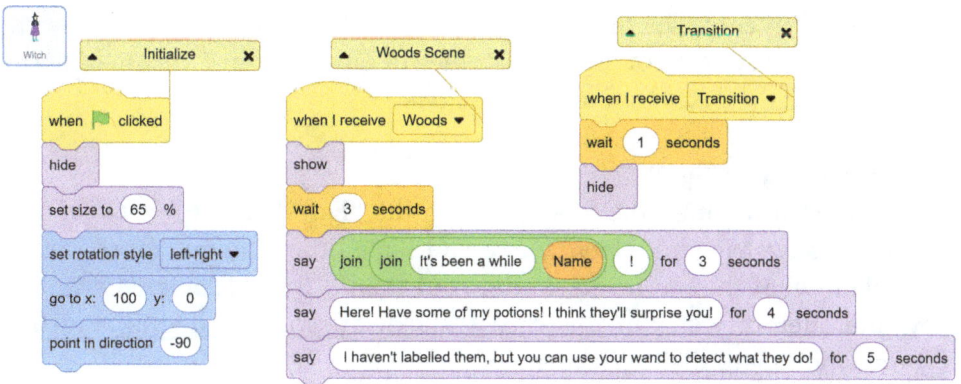

Step 12: The Gate and the Ogre
At the gate, the *wizard* encounters a threatening ogre blocking his way. At this point, we need to determine two possible scenarios, depending on the

previous gameplay. Either the *wizard* has the food from the farm or they have the potions from the *witch*. If we test ❶ our •*"HasFood"* variable, we'll know which of the two scenarios we have here.

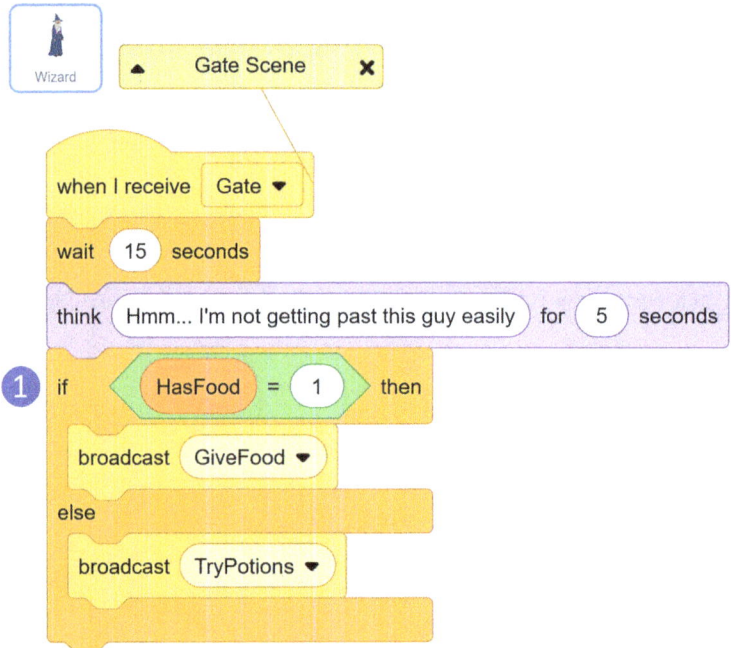

For this scene, we'll need *Frank* the ogre added to the game to set the final challenge. *Frank* needs the usual character starter code, but then we want two •*"Gate"* **events**. The first ❶ will set his dialogue in motion. The second ❷ will simply cycle through his costumes to give an angry, threatening display blocking the *wizard* so close to his goal.

Step 13: Feeding the Ogre

If the *wizard* has gotten the food *gift* from the *"Farm"*, the ❶ •*"GiveFood"* event will have triggered. Here we can add in a little dialogue for the *wizard* to explain what they're trying to do. And then a ❷ •*"FedOgre"* event for how the attempt went, in this case ❸ a delay, some dialogue, and then ❹ transitioning to the win screen.

At this point, we'll need to return to our *gift* sprite to finish up its code. We'll add a ❶ •*"GiveFood"* event that will position the *gift* under the *wizard* and make it visible. Then in the •**[When This Sprite Clicked]**, we'll fill in

Intermediate Project 2: Interactive Story ◆ 53

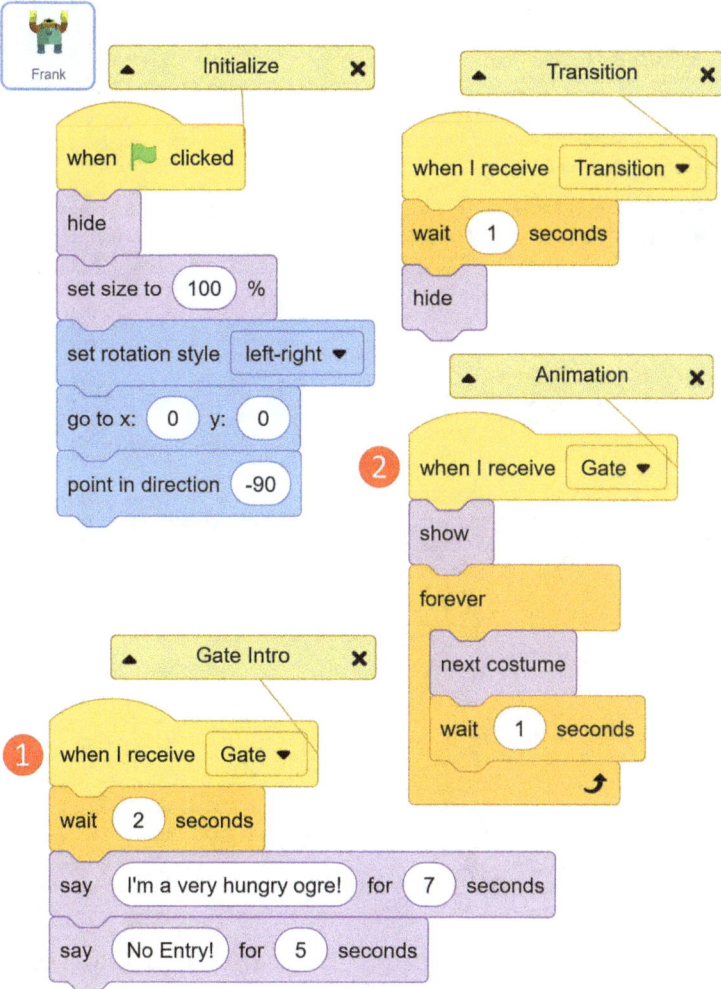

the ② •**Else**. This is, of course, the other scene that the *gift* can be clicked on. Here we add in another message, •*"FedOgre"*, that will make the ③ *wizard* react to the result of this action. The food will be passed to the ogre using a ④ •**Glide**; there will be a delay, then the *gift* moves back to the *wizard* with another ⑤ •**Glide**, then •**Hides**.

Lastly, we need to edit *Frank*'s code. We'll add in a ① •*"FedOgre"* event for them to react to. It turns out that *Frank* has about the same opinion of •*"Food"* as the player does! They ② stop their animation by stopping the other scripts in this sprite, then ③ *Frank* moves off-screen and ④ disappears. It might not have worked how they expected, but it did work!

54 ◆ Intermediate Project 2: Interactive Story

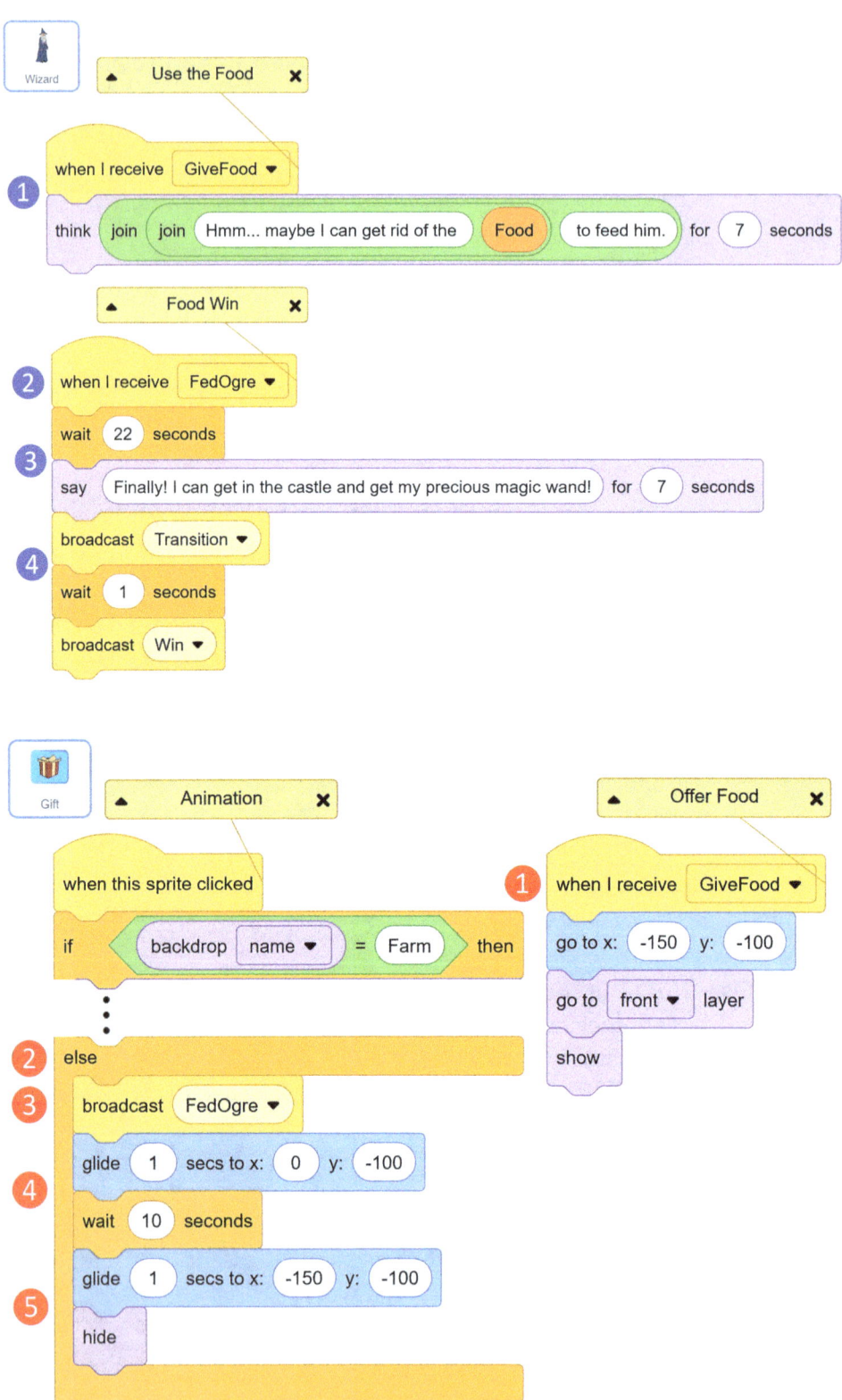

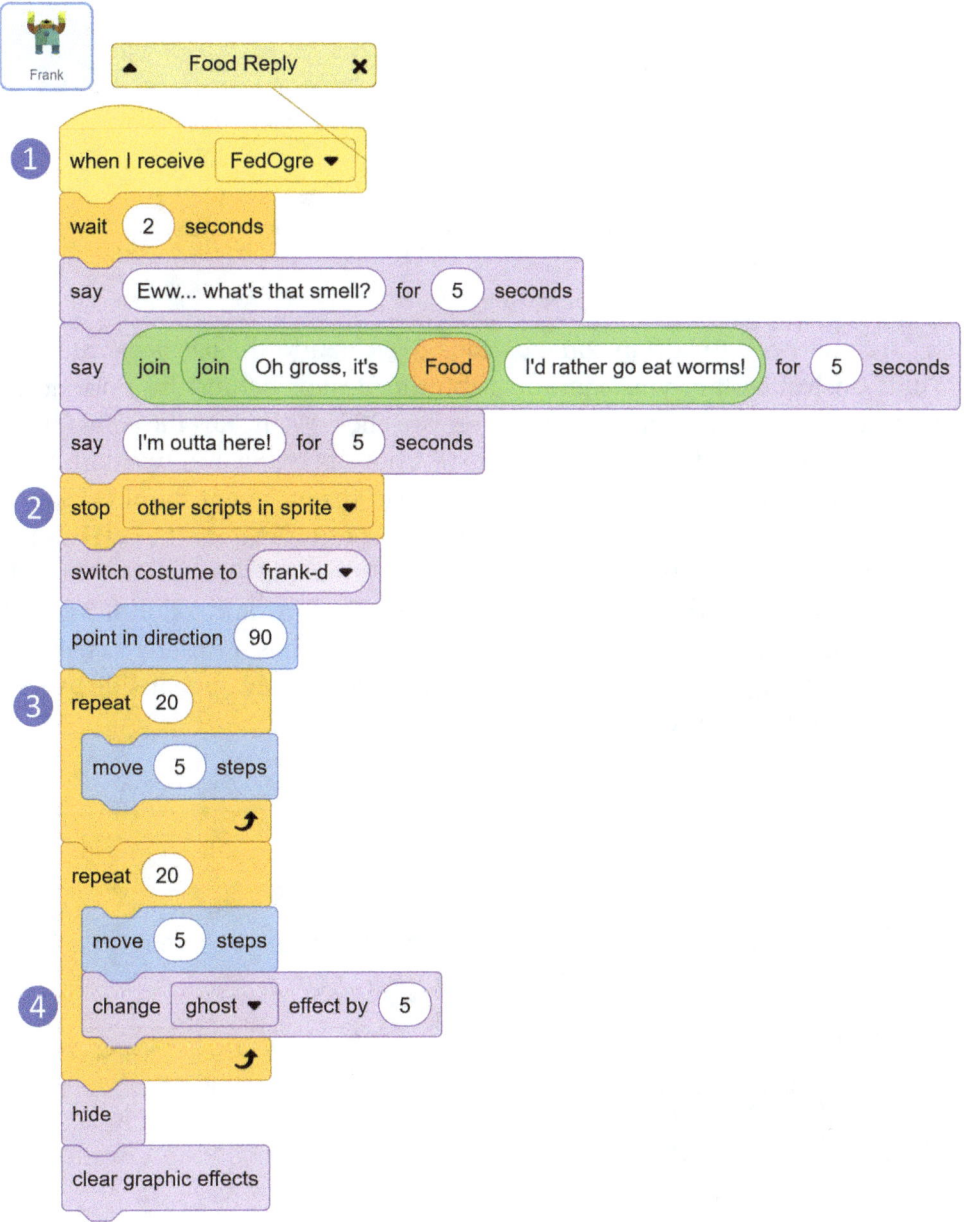

Step 14: Testing Potions

If the *wizard* went the woods route, they ended up getting potions from the *witch*. In this case, the ❶ •*"TryPotion"* event is triggered, so let's give the *wizard* a line explaining what they're thinking.

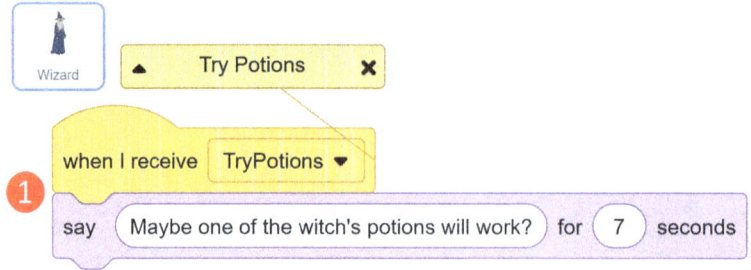

Now we need to add in some potions. We can add the *potions* sprite, and in the **Costumes** tab, you can add a button design behind the individual potion vials if you want to make them easy to click. We'll start the code for the potion by adding in the usual starter code from the *griffin* and tweak it. We'll need to add in a few events here. The ❶ •*"TryPotion"* event should •**[Show]** the potion. Each potion should have ❷ two potion events that •**[Hide]** it, and the ❸ •**[When This Sprite Clicked]** event runs the third. With that code complete, duplicate the potion object to make a **Potion2** and a **Potion3** object. In each, delete all but one of the costumes, so each has a different costume, and tweak the ❹ •**[Go To X:(#) Y:(#)]** coordinates so they appear side by side along the bottom of the screen.

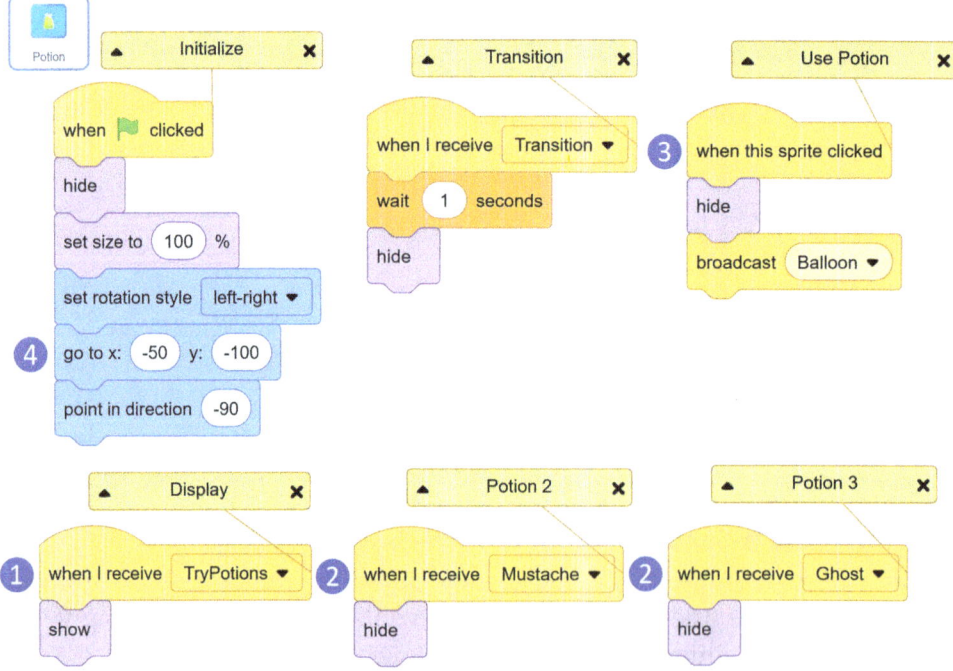

Intermediate Project 2: Interactive Story ◆ 57

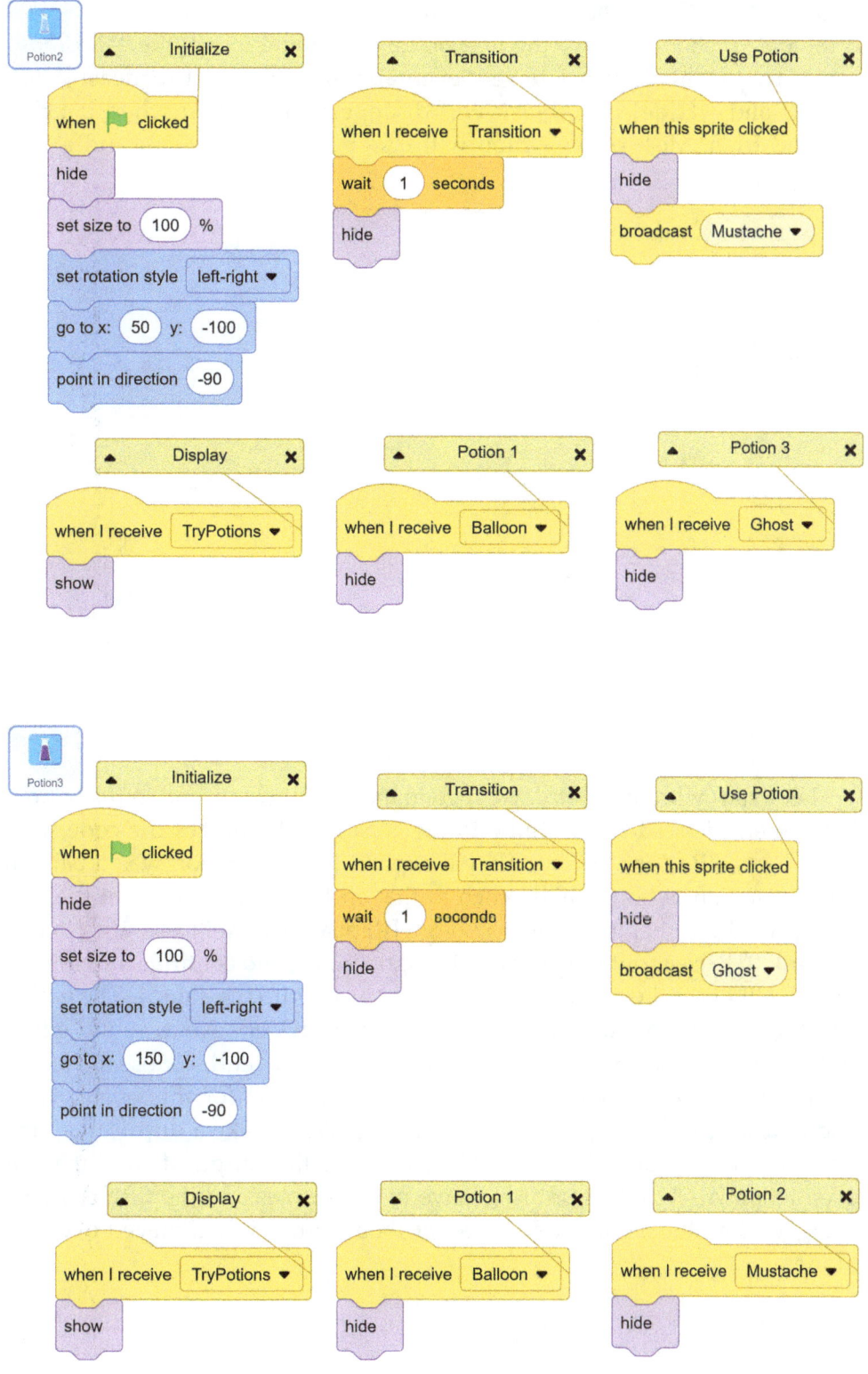

Step 15: Potion #1 – Balloon Transformation

If the player chooses to try the first potion, it's going to run the •*"Balloon"* event. The •**[When Sprite Clicked]** event in *potion* with a •**[Broadcast ["Balloon"]]** and •**[Hide]** code block will disappear this potion when it's chosen and tried. Most of the action will take place in the *wizard* and a new sprite – *cloud*.

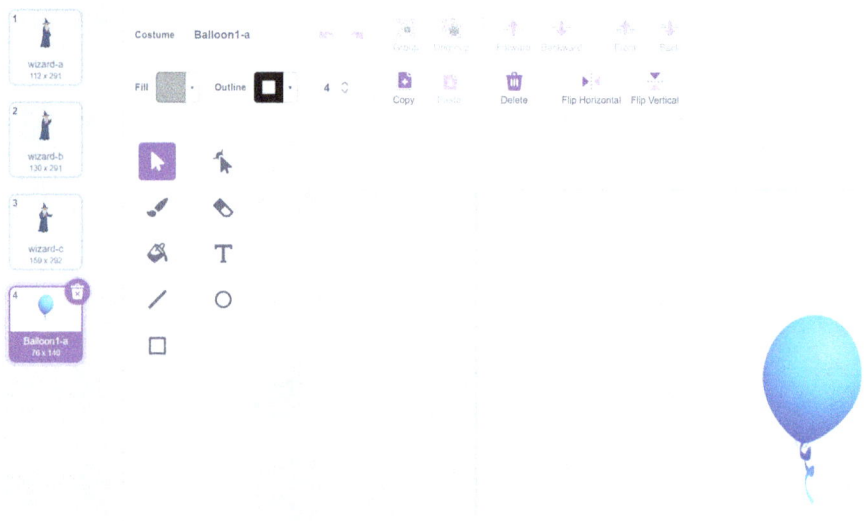

Let's start with the *wizard*. We'll go to the **Costume** tab, and in the bottom left corner, click **Add a Costume** to the *Wizard*: **Balloon**. For the code, they need a ❶ •*"Balloon"* event. They'll explain that the potion was a balloon transformation potion. We'll switch to that costume, then we point the *wizard* upward and begin a slow-moving •**[Repeat (#)]**. We'll use a ❷ blank •**[Think ()]** to clear out the previous thought bubble and continue moving up. Then we'll ghost the player as they disappear up into the clouds. The *wizard* drank the wrong potion! We'll add a ❸ •*"GameOver"* event, sound effect, •**[Hide]** the *Wizard*, and reset the graphic effects. A sad end so close to victory.

We're not done, though. Let's add another effect for this event. We'll add a *cloud* sprite and go to the Costume tab. We'll make a **duplicate** of the first costume and stretch it a bit taller. Then another **duplicate** and make it even taller. We'll do this until we have five costumes and the fifth costume is bigger than the *wizard* and can cover them completely if placed over top of them.

Intermediate Project 2: Interactive Story ◆ 59

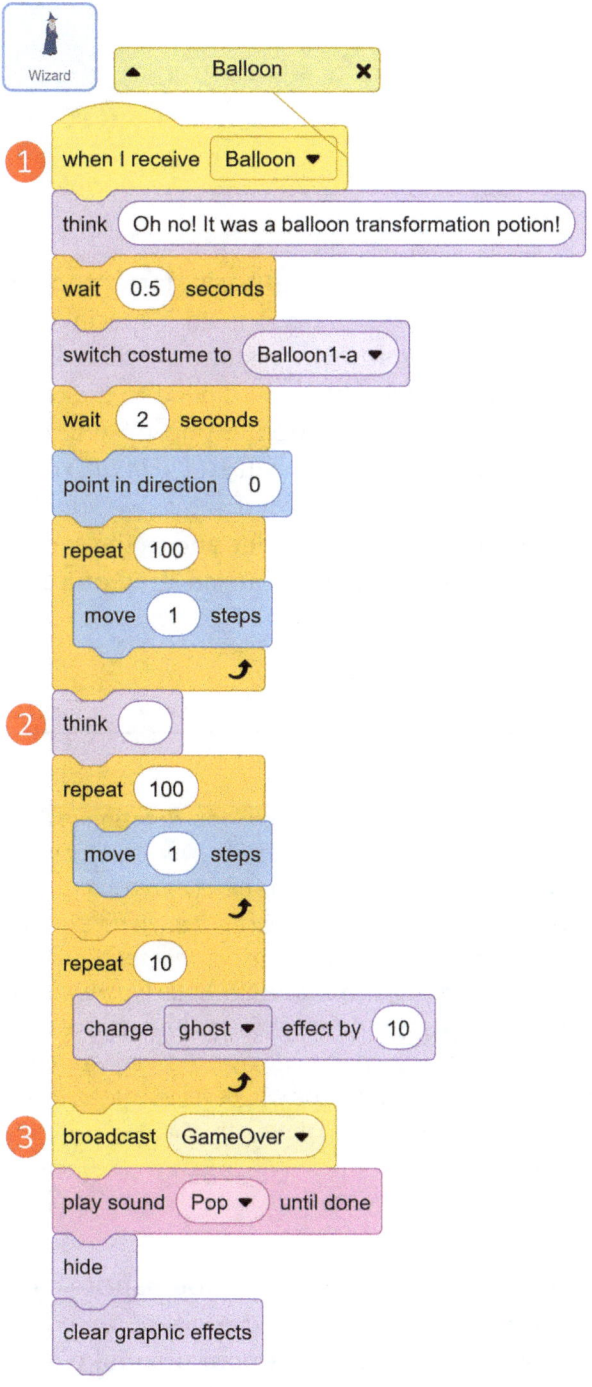

In the *cloud's* code, we'll need the ① basic starter code, again tweaked to its particular needs. Now we're going to need a special code block we haven't dealt with before, your very own custom code block. For this, go to the •My Blocks category. Like a variable, you'll need to **make a block** before we can use one. When you click the button, a pop-up will appear. You'll type in a name for your code block; we'll call it •*"CloudExplosion"*. Click OK. We now have our own code block. You'll see a ② •**[Define *"CloudExplosion"*]** block that you can attach code under, and in the code block view you'll see a new •**[*"CloudExplosion"*]** block you can drag out. First, let's •**[Define *"CloudExplosion"*]**.

> ### Functions
> *Custom code blocks are basically a concept called a **function** in other languages. They allow us to make a set sequence of code, like a recipe, that we'll be able to call whenever we like by placing our custom code block into code stacks. We define this function in one place and then can access, or call, it anywhere we want. Custom code blocks are object-specific, so if you create one in a sprite, it will only be available in that specific sprite. You can duplicate a custom code block into another sprite, though, if you do need to reuse the same function somewhere else. Breaking code into functions or scripts, also known as "scriptifying" or "encapsulating" code. You break it into smaller chunks and use those whenever you need. It's a great way to make code more readable, simpler, and more versatile.*

Our •*"CloudExplosion"* is to have the cloud sprite create an animation where the cloud grows rapidly to cover the *wizard*, then ghosts away. Much like a magician disappearing behind a poof of smoke, we use this animation to hide the changes to the *wizard* from the magic potions. We'll add a ③ •**[When I Receive [*"Balloon"*]]** event to our *cloud* and then attach our •**[*"CloudExplosion"*]** block underneath. Now, when the *wizard* uses *potion*, the *cloud* will poof in front of them and hide their transformation into a balloon.

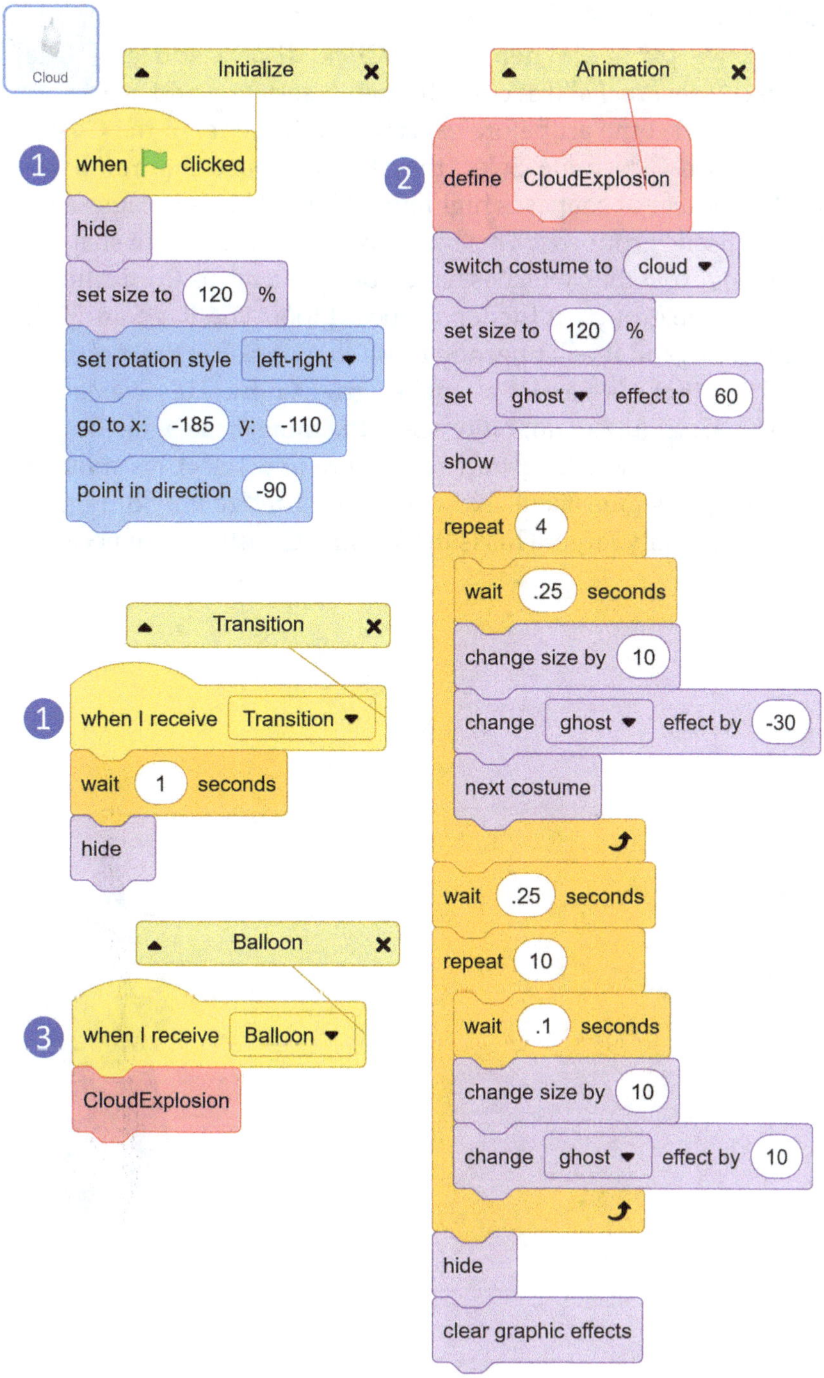

Intermediate Project 2: Interactive Story ◆ 61

Cloud

Initialize

① when 🏁 clicked
hide
set size to 120 %
set rotation style left-right
go to x: -185 y: -110
point in direction -90

Animation

② define CloudExplosion
switch costume to cloud
set size to 120 %
set ghost effect to 60
show
repeat 4
  wait .25 seconds
  change size by 10
  change ghost effect by -30
  next costume
wait .25 seconds
repeat 10
  wait .1 seconds
  change size by 10
  change ghost effect by 10
hide
clear graphic effects

Transition

① when I receive Transition
wait 1 seconds
hide

Balloon

③ when I receive Balloon
CloudExplosion

If you'd like to know more about animation and special effects, the Fireworks project in Book 1: Beginner offered some great techniques to know about. In Book 3: Advanced, we'll cover more character animation and deal with walk cycles in the Point-and-Click Adventure game.

Step 16: Potion #2 – Moustache Growth

In **potion 2** we put a •**[When Sprite Clicked]** and add a •**[Hide]** and a •**[Broadcast [***"Moustache"***]]** event to trigger its selection and use. In our *wizard*, we'll need to add some art before we can add the code. In the **Costumes** tab, select the **Wizard-C** costume. Select the *wizard*'s moustache. You'll see that the outline of a number of shapes is highlighted blue. This is because a number of shapes have been "grouped". This basically locks a number of shapes together so they all move together, which is handy but prevents us from editing just one component of the group. At the top of the art tools you'll see an ❶ **Ungroup** button. Click this and it will ungroup those shapes. Select the moustache again and you'll see that it's part of a smaller group now. This was a group in a group. So hit **Ungroup** again. Now select the moustache again and you'll see you've just got his face. **Ungroup** once more and select his moustache. We finally just have his moustache. Now we can ❷ resize it by dragging the corners out. Make it silly large. This is the **Wizard-C** costume we need.

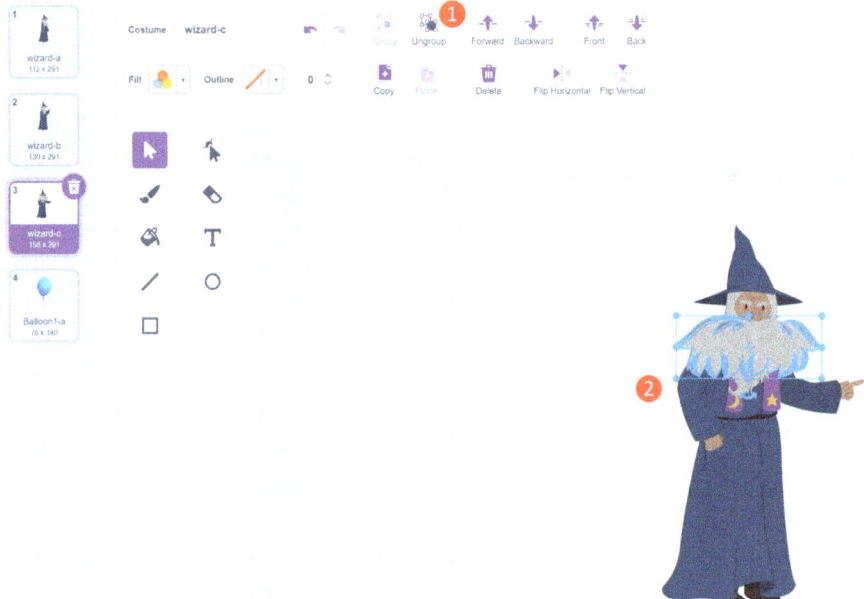

Intermediate Project 2: Interactive Story ◆ 63

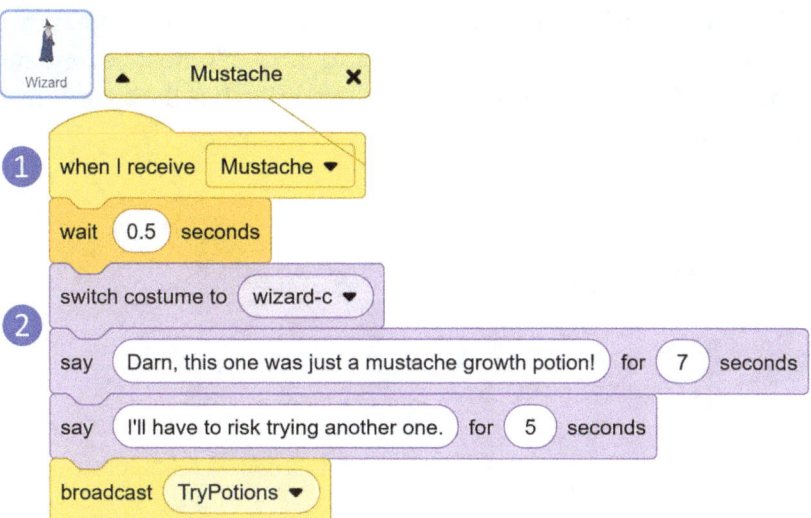

Go back to the code, and in a ① •"*Moustache*" event you can include a ② costume change to *Wizard-C* and add in some dialogue. This potion isn't final; it just makes the costume change and then lets the player make another selection. Now we can still use our cloud effect with this potion too. If you go to the *cloud*, you can add a ① •"*Moustache*" event and have it run the • ["*CloudExplosion*"] function there too.

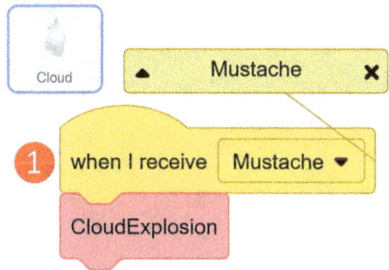

Step 17: Potion #3 – Ghost Transformation

The final potion, *potion 3*, has a •[**Broadcast** ["*Ghost*"] event when clicked. In the *wizard* we're going to have to add some more art. Go to the **Costumes** tab and click **Choose a Costume** and select the screaming **Ghost-C**. In our code we'll need a ① •"*Ghost*" event. Here our *wizard* will ② change costume to the *ghost*; we'll add some delays and dialogues and then ③ transition to the Win screen. In our *cloud* sprite, we'll add a ① •"**Ghost**" event and put a •[*"CloudExplosion"*] under it so the transformation has the appropriate magical "poof" animation.

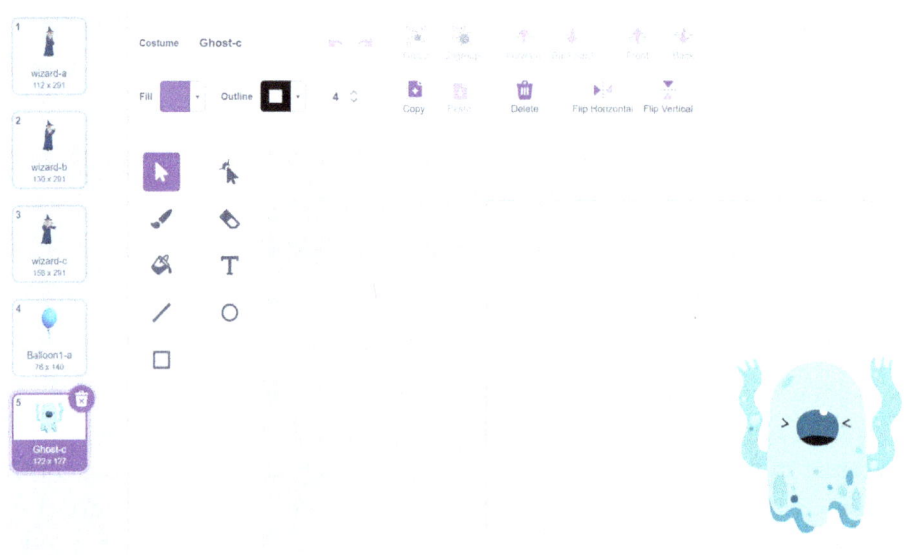

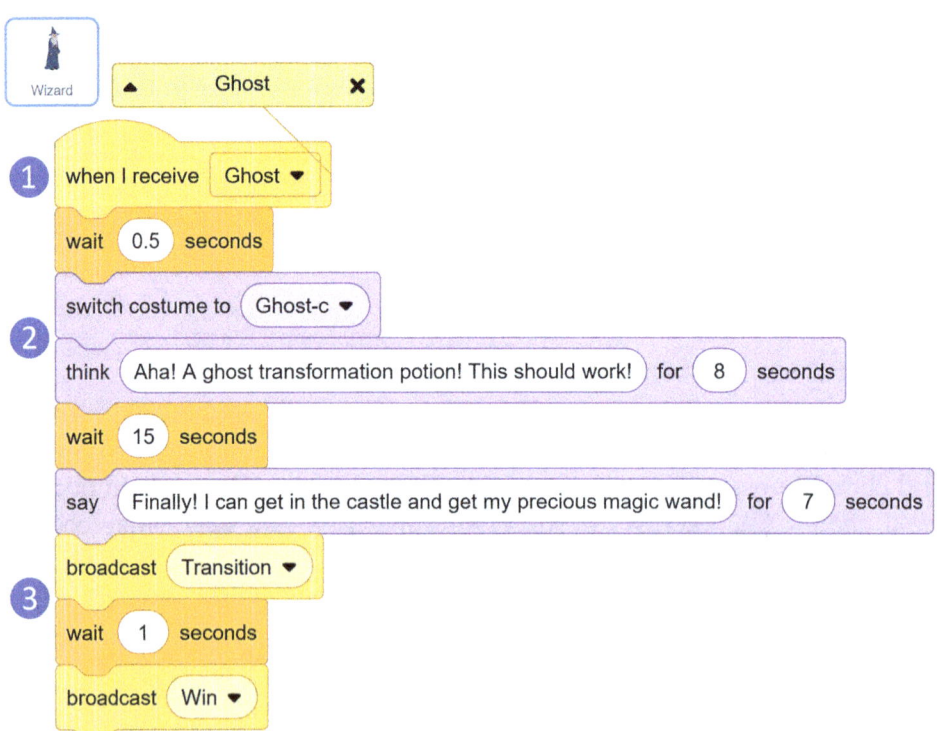

Intermediate Project 2: Interactive Story ◆ 65

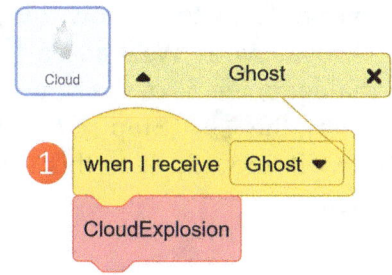
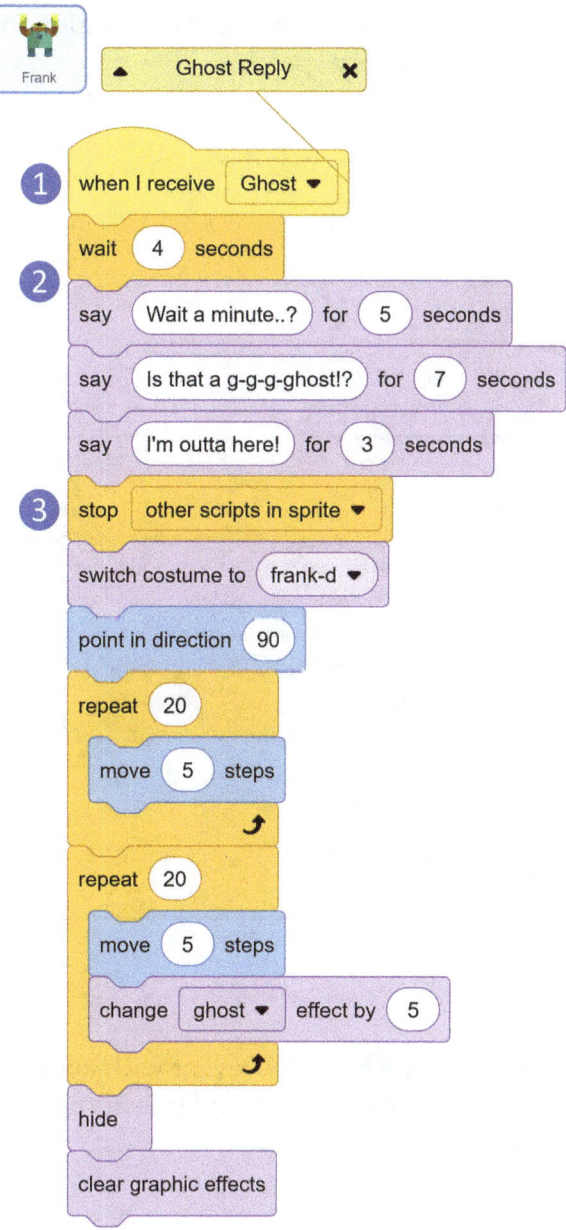

Now we can switch over to *Frank* to see how this potion has an effect on him. He'll need a ① •*"Ghost"* event. We'll add a ② delay and some dialogue; after the custom dialogue, we have the same sequence as the •*"Fed-Ogre"* stack. So you can copy the ③ •**[Stop [*Other Scrips In Sprite*]]** code block and everything below it and add it to this stack. We now have two solutions to the ogre problem.

Step 18: Win Screen

With the •*"Win"* event called, we can go to our *wizard* and add a ① •*"Win"* event. We'll add a little delay for the transition to fade to black, ② switch the *Wizard* to their normal state, then delay for the transition to fade back. Then ③ we'll give a little game victory speech.

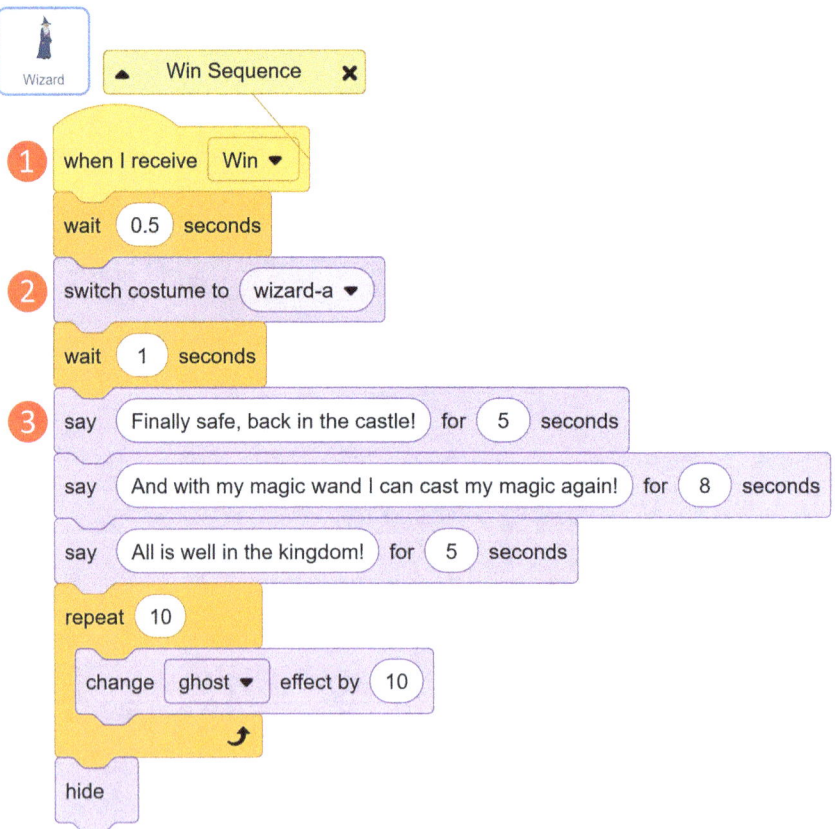

We'll add to our *cloud* a ① •*"Win"* event with another •[*"CloudExplosion"*] custom block. This'll have our *wizard* transform back to normal with the appropriate poof.

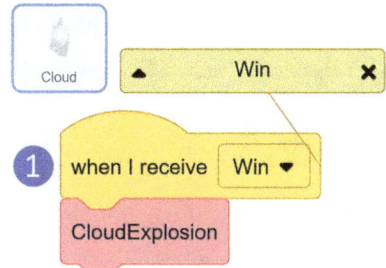

In our *fader* we'll have a ① •*"Win"* event. After a ② long delay, we'll send it to the back; this will allow our *goblin* to still be visible overtop of it, and ③ we'll decrease its ghost effect, making it fade to black.

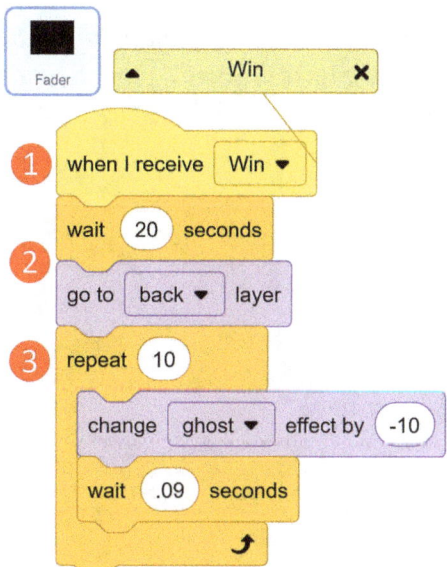

In our *goblin* we'll add a ① •*"Win"* event too. We'll add ② a long delay, then have them •[Show] and ③ congratulate the player on making it successfully through the story, then ④ end the game with a •[Stop [All]].

You've now got a sample interactive story project to help you and your students design their own interactive fiction with a whole range of techniques to work with!

Intermediate Project 2: Interactive Story

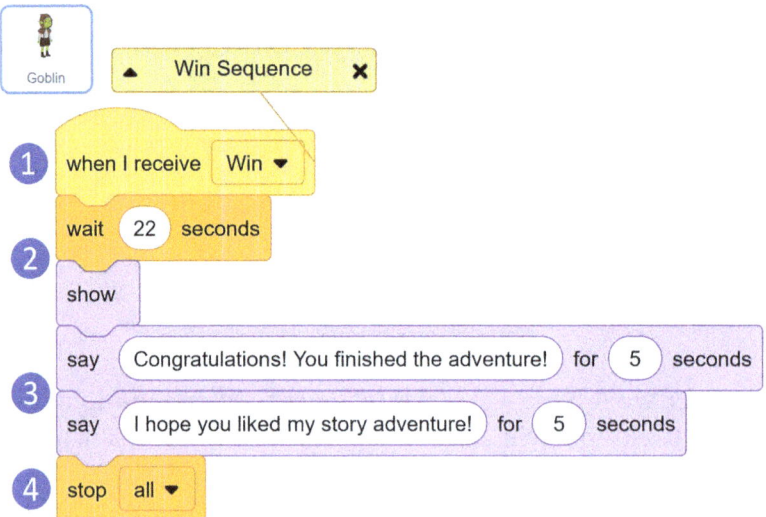

For working copies of this and every project in the book series, visit **www.massivelearning.net** *for direct links to Scratch projects, and to see our other projects and resources for coding education!*

7

Intermediate Project 3: Snowball Fight

What This Project Is

This game is a variation on the classic artillery game. The player controls a snowball-throwing reindeer trying to hit a snowman on the other side of the screen by controlling the angle and power of each throw. The snowman throws back a snowball each time the reindeer makes a throw. The player is encouraged to earn as many points as possible hitting the snowman or stars before the snowman hits them back with a snowball, ending the game. This is a great way to practice our physics modelling in Scratch and should take about an hour to complete for an adult.

What We're Learning with It

This game is all about ballistics, so we're going to learn some methods for creating game physics. We'll make an arcing motion using •Variables to model ballistic curves. User input will drive variable settings as well as interface elements to convey their choices. Lastly, we'll add in a number of game conventions, like indicators/readouts, turn segments, as well as Start and Game Over screens. Here we really start learning the polish of game development, with more focus on user interface and control, adding another level to our projects and helping them stand out.

Player Turns. How to divide our game into player turns, alternating between the player and the computer opponent throwing snowballs.
Object Collisions. Test for objects colliding to check for snowball hits on both reindeer and snowman, bonus scoring objects, as well as blocking objects.
Start Screens. Add in a Start screen for our game, providing a title image before jumping into the action.
Game States/State Machines. Learn to divide our game into different states, letting our game operate differently, whether it's on the Start screen, running the game (in either player or computer turn), or Game Over.
Computer Opponents. We'll add in a computer opponent that will challenge the player with the risk of losing.
Gravity. How to incorporate a gravity effect to pull the snowball down to the ground.
Physics Modelling. Two ways, one using •Variables, will be used to model out a simple curve of a thrown snowball.

Building It

Step 0: Create Your New Project

Make sure you're logged in to Scratch, then click **Create** to begin a new project! Since we won't be using it, we can delete the Scratch Cat sprite by clicking on the trash bin on that sprite's thumbnail in the **Sprite Listing**.

Step 1: An Arcing Snowball

Start by setting the background to the **Arctic**, a suitable winter scene for our snowball fight. Then add the *baseball* sprite (or add your own custom sprite) to act as our snowball, though feel free to make your own art. If using the *baseball*, I'd recommend changing the **Size** property to 25. You want it fairly small, so accuracy is an issue. Name the sprite "*Snowball*".

Our snowball needs to be able to arc through the air to reach its destination. We'll start by building out some of the basic properties of its motion. First, we'll use clones to make copies of the snowball when launched. Start with a ❶ •**[When This Sprite Clicked]** to •**[Create a Clone of [*Myself*]]**. We'll have our original *snowball* as launcher for the clone snowballs. Position it on the left-hand side of the screen about midway to the top or a little above.

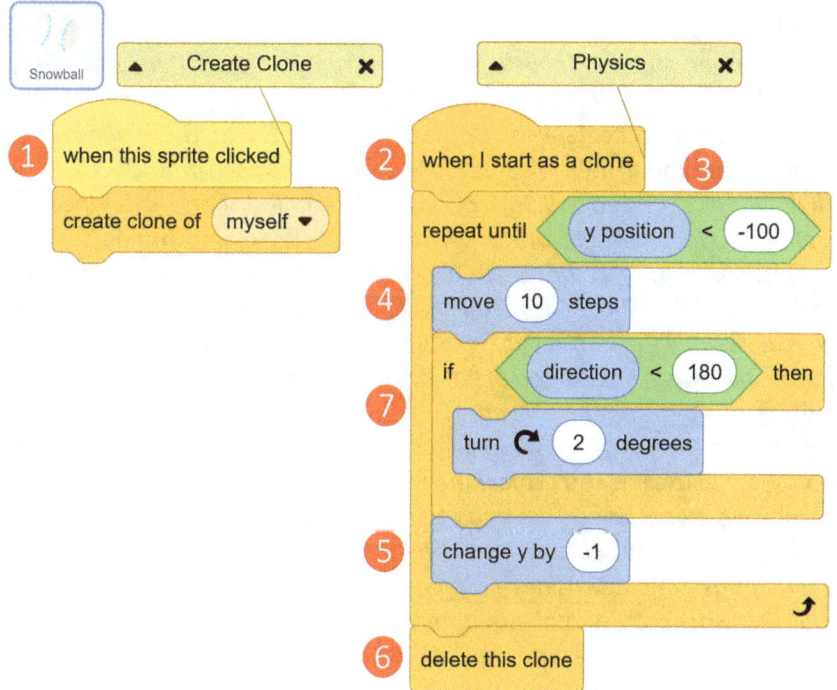

To throw the snowballs, have them trigger with a ❷ •**[When I Start As A Clone]**. Use the code block •**[Repeat Until <condition>]**. This is just like an infinite •**Repeat** loop, but a condition can be added to stop it from happening. In this case, it will end if the *snowball* falls to ❸ •**<•(Y Position) < (-100)>**, basically if the *snowball* hits the ground.

Within our •**[Repeat]**, let's start with a ❹ •**[Move (10) Steps]**. Our snowball now flies right but never drops. Add in a ❺ •**[Change Y by (-1)]**; now it very slowly drops to the ground over time. With a ●[Delete This Clone] under our ●[Repeat], our clones it will self-destruct when it hits the ground. Our *snowball* is flying and dropping, but it's actually just a diagonal line, so next let's make it a curve!

By adding a ●[Turn Right (2) Degrees] block, our *snowball* begins to arc. It starts at 90, facing right, but now curves its path down toward the ground. To improve this, put it inside an ●[If] so it only turns right *if* the ●(Direction) isn't already straight down. Gravity should pull it to the ground, but not past it, so this ●[If] will make sure it ends arcing in a straight downward plummet. We've got a pretty good start to our snowball-throwing. Try moving the original *snowball* around and see how the motion looks from different heights.

> **Solutions to Fit**
>
> *Some of you with more of a physics background may be wondering about how we're modelling arc motion here. We aren't using the kind of dynamically adjusting horizontal and vertical speeds that you might be used to. This is one of the fun parts of programming; there are lots of ways to solve a given problem. Classically, Earth gravity is measured as 9.8m/s per second, meaning, it gets faster the longer you fall; it's a dynamic value rather than a set value, like our ●[Change Y by (-1)]. Here it just doesn't matter that much, so we use a very simple system. We could use a dynamic calculation to have the motion change, but here we use a simple cheat to get nearly the same effect – rotating our snowball. An arc should have a nice parabolic curve, but instead of having a complex calculation like gravity, we just rotate, and by using that, we introduce arc motion without the fuss of trigonometry. This is because we're just making a silly game; if we were making a physics demo, we could use the exact calculations. The same platform, same concept, but different ways to imagine things and code them. Don't be surprised when your students come up with solutions you never thought of. Each will have its own reasons, upsides, and downsides – explore them. Understanding the consequence of choices is what professional programming is all about. And don't worry, we'll give you examples of some more dynamic modelling in step 3, as well as in Book 3's Platformer game.*

Step 2: Throwing Angle

An artillery game is all about getting to choose the angle and power of your throw. Add in an ●*"Angle"* variable to let players take aim. The ●*"Angle"* variable needs a slider with a value range of 1–100 (the default). To set it to a slider, right-click on the variable display on the Stage Window, then right-click on it again to set the slider range. You might wonder why you'd want to throw straight up or even away from the target (90–100 degrees), but I'll touch on that later.

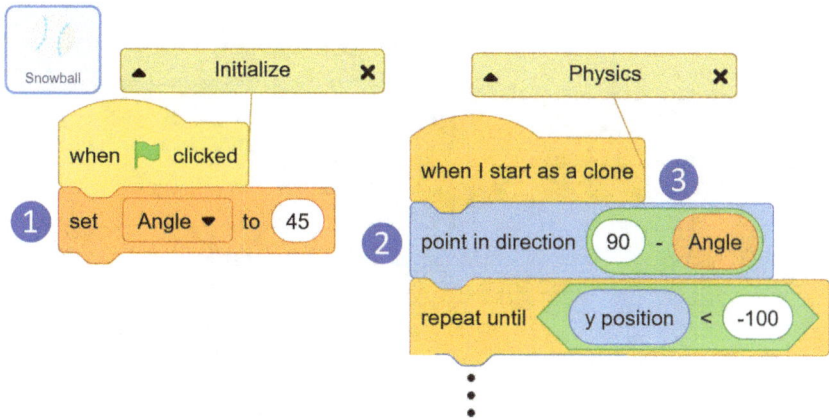

Set the ①•*"Angle"* variable to 45 as the default in a •[When ▷ Clicked] event. Then, add a ②•(Point In Direction (#)) code block to make sure our aiming has an effect when a clone is generated. In this case, we need to transform the angle of 1–100 to what we really need, in this case, ③•((90) – •("Angle")). Pointing right is the default, and each degree of •*"Angle"* raises the shot above the horizon. Try it out! It can be fun seeing the different arcs.

Step 3: Throwing Power

Now that we've got our throw's •*"Angle"*, let's add in a dynamic •*"Power"* rating. Add a •Variable named •*"Power"*, make it a slider 1–100 range, and ① set it to 50 in our •[When ▷ Clicked] event. The exact •*"Power"* level of a throw is going to dynamically change through the course of the arc, so we need a second variable to hold this dynamic value. For that, make a hidden (uncheck it in the list so it doesn't display in-game) variable called •*"PowerIs"* to hold this actively changing number. Then, add in a ② •[Set (*"PowerIs"*) to (*"Power"*)] code block combo under our •[Point in Direction •((90) – •(*"Angle"*))] combo. This will start our new snowball at the •*"Power"* selected.

Next, we need to change our •[Move (#) Steps] code block. Here ③ we'll add in a •((0)/(0)) (Division) code block and use our •*"PowerIs"* variable as the first value, then a scaling factor to divide it. Let's use this to adjust the step size to our •*"Power"* rating. But why do we need •*"PowerIs"* instead of just using •*"Power"*? Because we're going to change it!

Under our •[Move (#) Steps], add a set •*"PowerIs"* code block. Here ④, we won't set it to a specific number but rather scale it to represent air

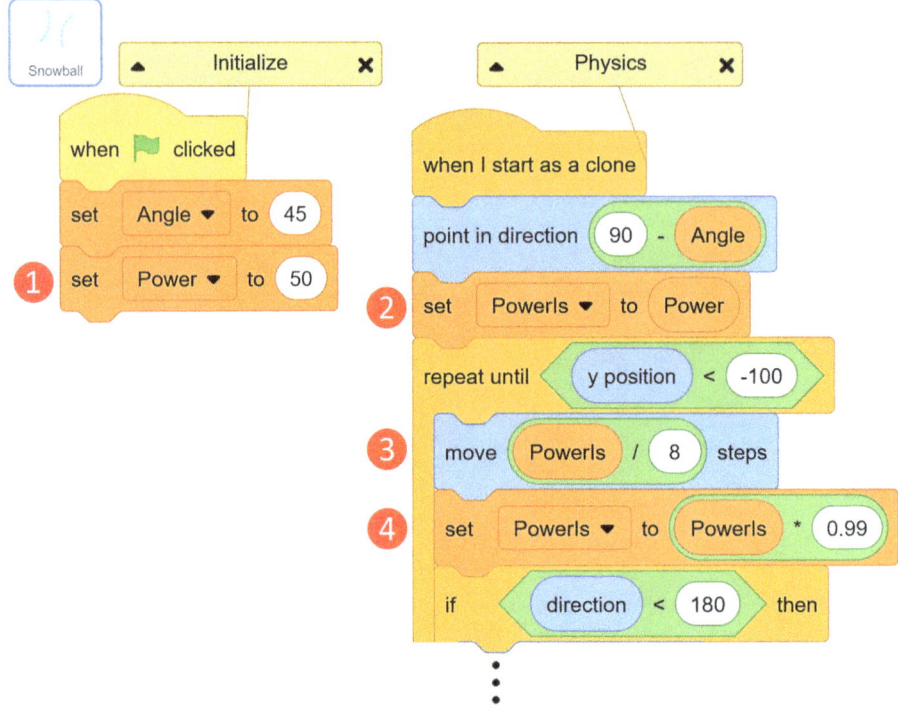

resistance. We'll use ●(●("*PowerIs*") * (.99)). It may not seem like much, but remember that runs 30 times a second. Our snowball will now slow down as it flies through the air. You can try different resistance factors to see how strongly this small change adds up.

Step 4: Aiming
Our ●"*Angle*" and ●"*Power*" are now dynamic, but let's add a visual representation of those values to help the player understand what the selections mean. Add an *aimer* object by creating a blank sprite. In the **Costumes** tab, ① make a circle, then choose the ② **Reshape tool**. A circle is made of four points. Grab the ③ left-hand point and drag it to the left about three or four times the width of the circle. Next, above the drawing you'll see ④ two options when a point is selected – **curved** or **pointed**. With the **Circle tool**, all points start on the **curved** option. On the **point** you've dragged out, change it to **pointed**. You'll see how instead of a rounded line this point now connects to the others with straight sharp lines. We now have a comet-shaped object. ⑤ Choose transparency for the outline (the square with the red diagonal line through it), and for the ⑥ fill, use a gradient left-to-right, with left being transparent, and right being bright red. It should look like a ●"*Power*" gauge! So it will line up correctly with our reindeer player later, move the shape so that the pointed transparent tip lines up with the cross, or reticle, in the ⑦ centre of the art workspace.

Intermediate Project 3: Snowball Fight ◆ 75

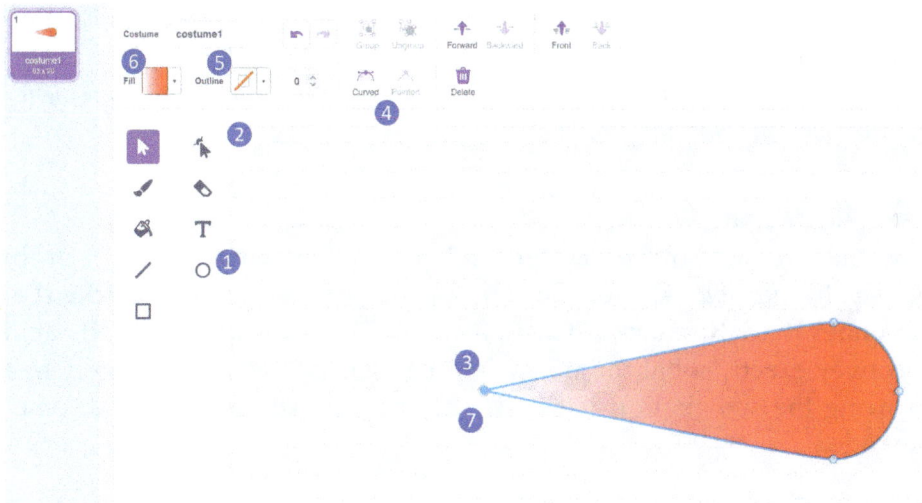

For its coding, use a ① •[When ▷ Clicked] event and •[Forever] loop. Inside, have it ② •[Go To [*Snowball*]] (this will move it to the original snowball, not a clone) and then [Show]. Next, we adjust it to reflect the current angle and power settings so it can be an accurate indicator gauge for the player. To do this, we'll add in ③ •[Point In Direction •((90) − •("*Angle*"))], and ④ •[Set Size To (•("*Power*") +(30)]. Adjust the number added to the **size** to make sense for your drawing; it may need a different number than 30 in the formula, depending on how large or small you made it in the Costumes tab, but we want to make sure not to shrink the *aimer* below a visible and clickable size. Your *aimer* now accurately shows the •"*Angle*" and •"*Power*" of the throw!

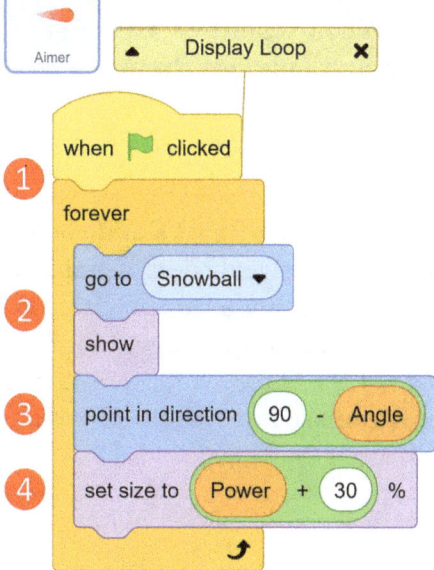

It's a lot of fun tossing snowballs, but if you need a simpler example of some object motion, you can check out the Fireworks project in Book 1: Beginner. You can learn about guide objects, a great technique for creating interesting movement, in Book 3's Scrolling Shooter project.

Step 5: Throw Event

Now that we have our *aimer*, let's adjust some of what we've set up previously. In the *aimer* sprite, add a ① •[When This Sprite Is Clicked] and put under it a •[Broadcast [Throw]] event. This new *aimer*, with its ability to convey exactly what the throw will be, makes a good user interface for throwing the snowball. If we switch to the *snowball*, let's change some of our code.

On our *snowball* sprite, ditch our •[When This Sprite Clicked]. Instead, add in a ① •[When I Receive ["*Throw*"]] event. We'll place our •[Create Clone of [*Myself*]] here. Now, our *aimer* (when clicked) will spawn the thrown *snowball*. With that new control in place, put a ② •[Hide] block under •[When ▷ Clicked] and add a ③ •[Show] block and a •[Go To [*Front*] Layer] to our •[When I Start as a Clone]. This way, our original *snowball* remains hidden and only our thrown *snowballs* become visible in the game!

Step 6: Our Target

Now that we can throw our *snowballs*, let's add something to aim it at. Add the *snowman* as a sprite. Right now, we don't need any code for it, so they're just going to be there as a target we can practice on. We can reduce their **size** property to 65, making it a harder target to hit. Position them at X: 150, Y: -100, on the right side of the Stage Window. In the **Costumes** tab, you can ① **Flip Horizontal** the sprite so it faces to the left and position it ② to stand on the reticle.

Intermediate Project 3: Snowball Fight ◆ 77

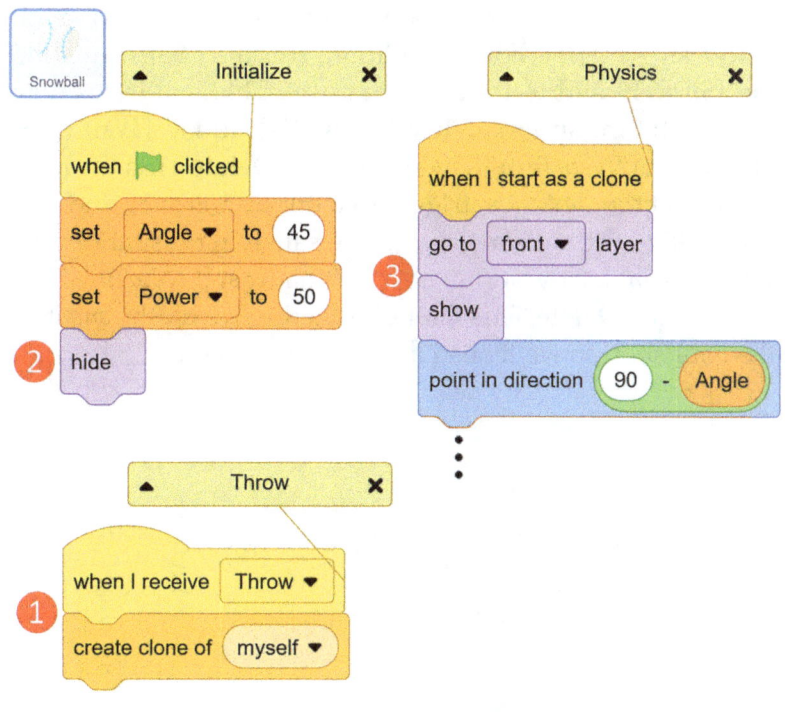

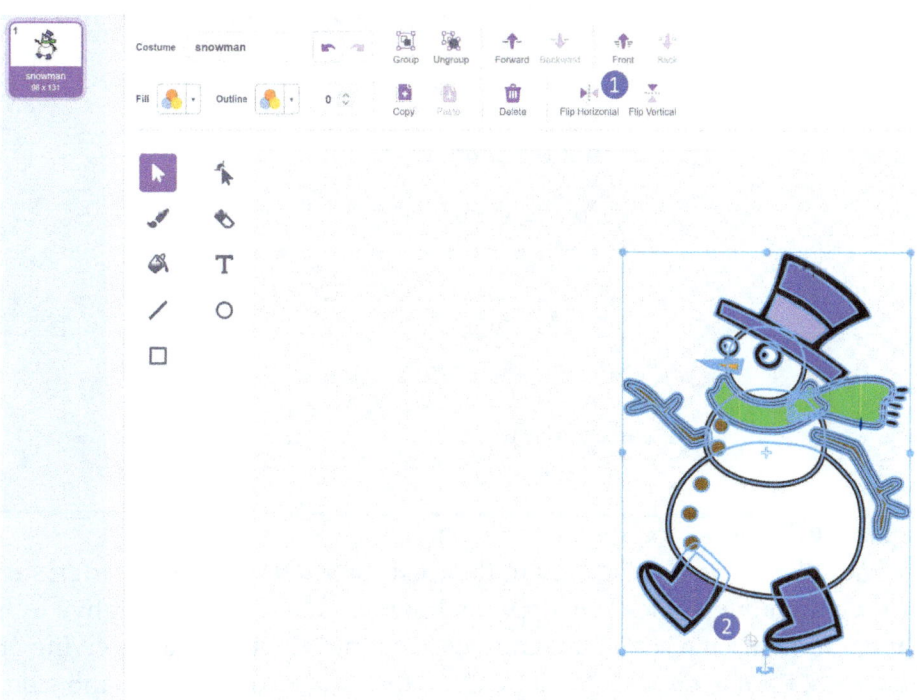

Our *snowball* now needs some code to detect the snowman. In our *//Physics* stack, add an ❶ •[If]. For a condition, use the •Sensing code block •<Touching (*Sprite*)> and test for the *snowman*. This will let us determine if the *snowball* has hit the *snowman*. Inside this •[If], add a •[Wait (2) Seconds], a new •[Broadcast ["*NewTurn*"]] event, and a •[Delete This Clone]. This will let the *snowball* hit the *snowman*, trigger that our turn is over, and delete itself, while also getting a little delay, giving the satisfying visual of our *snowball* hitting the *snowman*. Lastly, add in a ❷ •[Broadcast ["*NewTurn*"]] event just above our bottom •[Delete This Clone]. Whether we hit the *snowman* or not, we've taken our shot, so we'll need a new turn to begin.

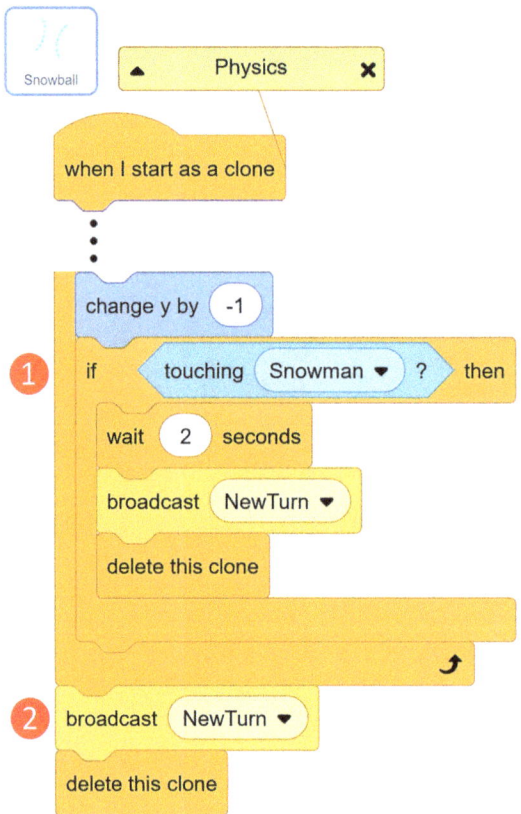

Step 7: The Player

Let's add the *reindeer* sprite, and in the **Costumes** tab, you'll want to position it to stand on the reticle. This happy fellow will be our representative of the player, throwing the snowball, and possibly ending up on the receiving end of one too! Resize it to 65%, then position the *reindeer* on the left-hand side of the Stage Window at the same height as the *snowman* (X: -150, Y: -100).

Intermediate Project 3: Snowball Fight ♦ 79

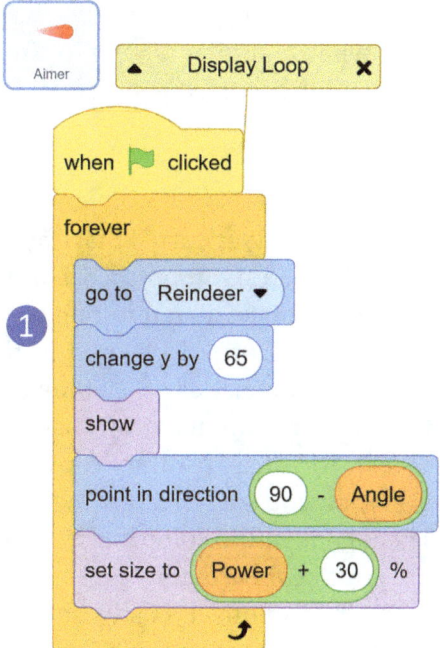

With the *reindeer* in play, we want to adjust our *aimer*. We can set it to ❶ •[Go To [*Reindeer*]] in the •[Forever] loop. This will help make sure our sense of •"*Angle*" is accurate. However, the centre of the *reindeer* is fairly low. So add a •[Change Y by (65)] just below the •[Go To (*Reindeer*)]. This will position our *aimer* up higher, basically centred on our *reindeer's* nose, making it nice and visible to the player and a bit more sensible that our *reindeer* throws from shoulder height.

We need to make a similar change in our *snowball*. Switch to it, and in the •*"Throw"* **event**; we want a similar ① •**[Go To [***Reindeer***]]** and then •**[Change Y by (65)]**. This will ensure the *snowballs* launch from the same point of origin as the *aimer*.

Step 8: The Turn Sign

The next step is adding in our player turns. Start by adding a blank sprite and naming it "TurnSign". In the **Costume** tab, we'll need two costumes for this sprite. Using the **Text tool**, make a sign in one costume saying *"Player's Turn"* and the other *"Computer's Turn"*. To keep it in theme, I suggest using the **Marker** font and two different shades of icy blue to differentiate the two. Next, we'll employ a technique to add in a drop shadow to our text using the arrow **Selection tool**. Next, for each costume, copy and paste the text (you can use Ctrl + C/Ctrl + V, or you can click on the Copy and Paste buttons above the drawing). Then, change the colour of the text to a darker version. Using the arrow keys on your keyboard, move it so that it's just to the bottom and left of the text with a bit of an offset. Lastly, above the drawing, click on the **Back** button to push the copy down behind the original text. This should give your text a nice shadow effect. It doesn't just look nicer, but it also helps it stand out from the background more, making it easier to read. Remember to do the same process for both costumes. Make sure the position for the sprite is X: 0, Y: 0, so it looks centred.

Finally, add the code for the *TurnSign*. ① •**[When I Receive [***"New-Turn"***]]** will run each new turn that's called. Add a •**[Hide]**, •**[Clear Graphical Effects]**, and ② •**[Switch Costume To [***costume***]]** with a new hidden (unchecked) •Variable named •*"WhoseTurn"*. This will allow us to reset the sign and make sure it displays the correct turn (player or computer). Our new •*"WhoseTurn"* variable will help us keep track of whose turn it is and make sure the code behaves for that condition, so nobody moves out of turn. We'll use an integer to track the turns, which here will be used to set the costume number of the sign, allow us to keep track of turns and costumes in one go. Add a ③ •[Show] and a •[Wait (1) Second] so players have some time to read the sign, and then use a ④ •[Repeat (20)], •[Wait (.05) Seconds], and • [Change [Ghost] by (5)] to make the sign fade away over 1 second. We'll end the stack with a ⑤ •[Hide]. We've now successfully indicated whose turn it is and then faded away so the game can resume.

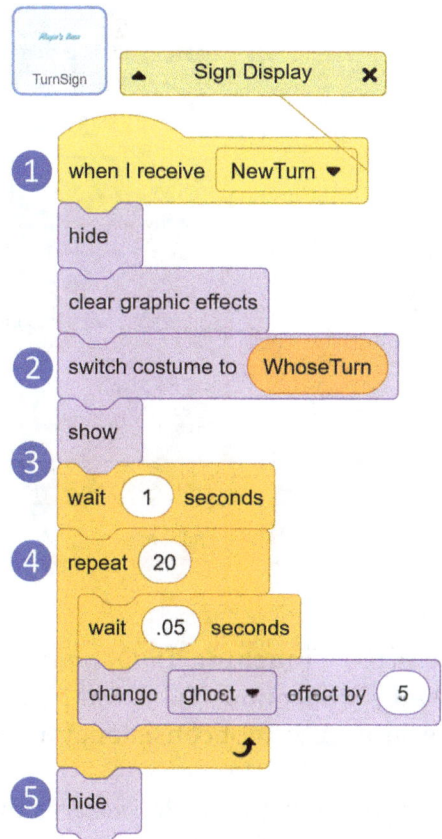

Step 9: Player Turns

Now let's make a system to integrate our signs! Starting with the *snowball*, we'll make a custom code block by going to •**My Blocks** and clicking **Make a Block**. Call our new code block •*"EndTurn"*. Once you've added that, you can call ① •*"EndTurn"* at the bottom of the •[When I Start As Clone] and move the ② •[Broadcast] and •[Delete] to the •[Define *"EndTurn"*]. But that's not quite enough. We want to make sure we switch whose turn it is when we end a turn. For that, add under the •[Define *"EndTurn"*] a ③ •[Set [*"WhoseTurn"*] to (*0*)], •[Wait (*2*) Seconds], and then a •[Set [*"WhoseTurn"*] to (*2*)].

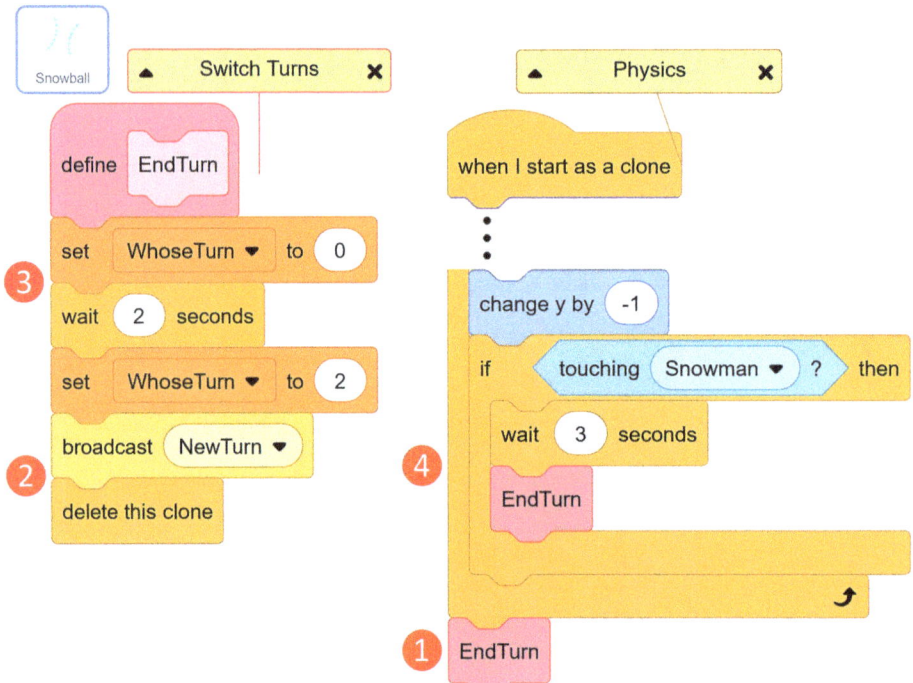

Also, change the •[If •<Touching [*Snowman*]?> Then]. Switch its contents to a ④ •[Wait (*3*) Seconds], followed by an •[*"EndTurn"*] code block. This will make sure we've got consistency, no matter how the turn ends.

Lastly, switch over to our *aimer*. Since the *aimer* is the way players throw their snowball, we want to prevent them from being able to throw when it isn't their turn. In the •[Forever] loop, add an ① •[If •<•(*"WhoseTurn"*) = (*1*)> Else]. This •If will determine if it is the player's turn or not. In the top section of the •[If], let's move in our existing code. In the bottom section •(Else), the one that runs if the condition isn't true (so it's the computer's turn), add a ② •[Hide] code block. This will mean the *aimer* disappears when it isn't the player's turn, and they won't be able to click it!

We can also add above the •[Forever] loop a ③ •[Set [*"WhoseTurn"*] to (*1*)] so the game will begin with a player turn. Now, one last tweak – we'll

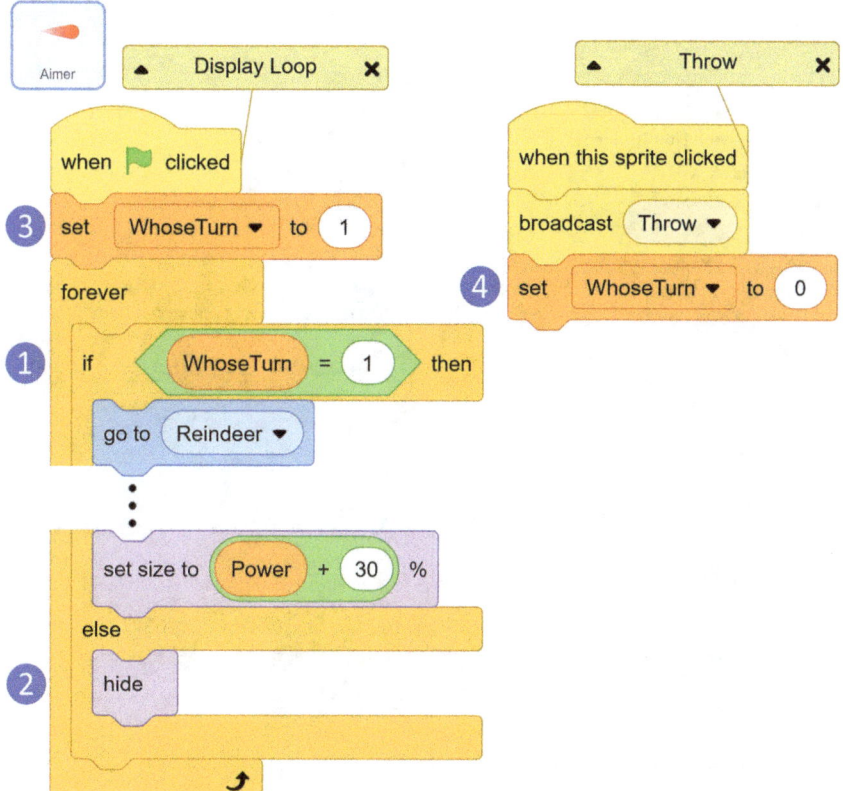

add a ④ •[Set ["*WhoseTurn*"] to (*0*)] under the •[Broadcast ("*Throw*")]. As soon as the player starts to throw, it ends their turn so they can't try throwing multiple snowballs in one turn! •"*WhoseTurn*"=0 is neither the *reindeer*'s nor the *snowman*'s turn, so it acts as a holding pattern to let the turn resolve before changing.

Step 10: The Snowman Strikes Back!

Now that we've got the player turns working, we need the *snowman* able to throw a snowball. First, duplicate our *snowball* object and name it *Snowball2*. You may want to alter the art a little to tell them apart easier.

Snowball2's code needs a few changes to work, and we'll start by switching the •[When I Receive ["*Throw*"] to a ① •[When I Receive ["*New-Turn*"]] event. Add an ② •[If •<•("*WhoseTurn*") = (*2*)> Then] and move the rest of the code inside that •If. That will make sure *snowballs* are not thrown out of turn. We also need to change ③ *reindeer* to *snowman* in the •[Go To (*sprite*)].

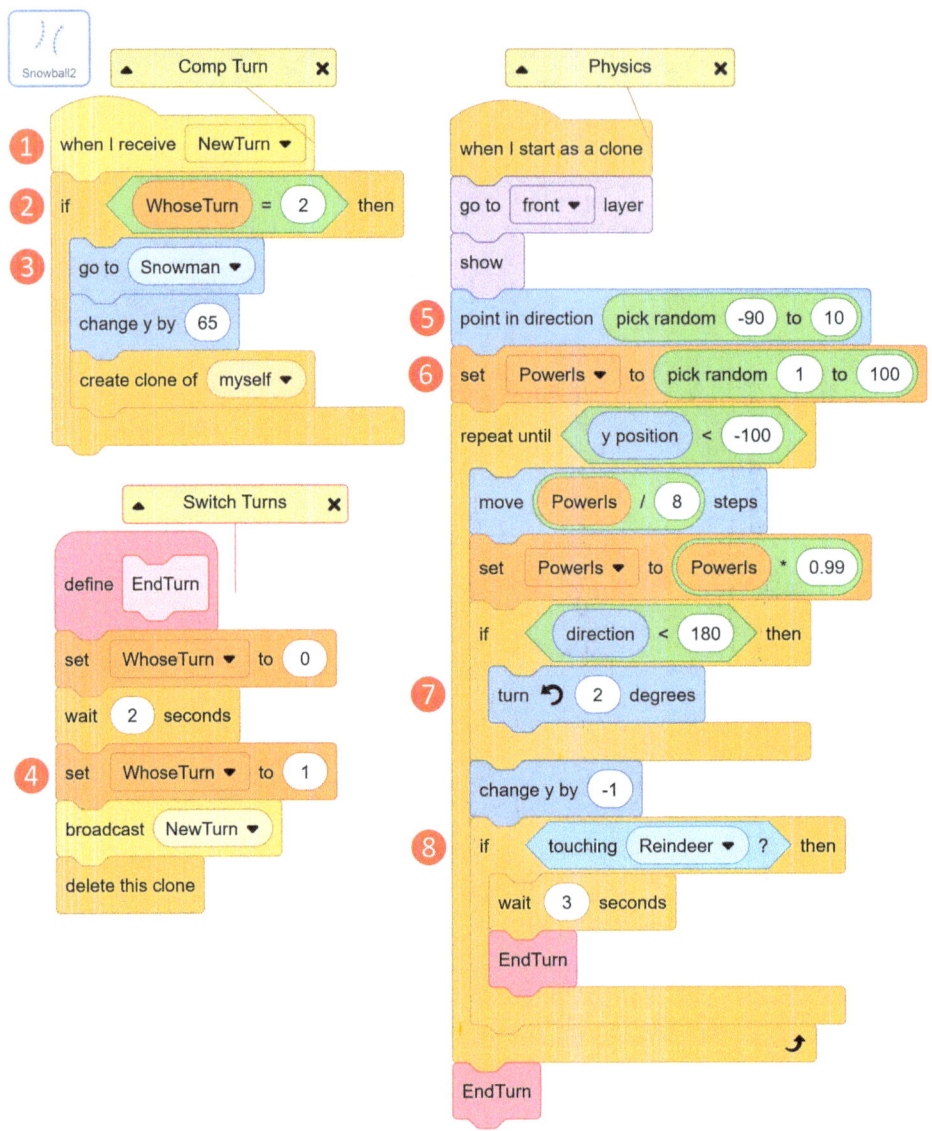

In the •[**Define** *"EndTurn"*], change the second ④ •[**Set** [*"WhoseTurn"*] **to (1)**] so that the *snowman* passes control back to the player.

In the •[**When I Start As a Clone**], we need to make a few changes. We don't want to refer to the •*"Angle"* variable, so replace the ⑤ •((90)- •("Angle")) with •(**Pick Random (-90) to (10)**), making the *snowman* throw randomly. We need negative numbers to aim toward the left instead of the positive numbers that aim right. Change the ⑥ •[**Set** [*"PowerIs"*]] from •*"Power"* over to another •(**Pick Random (1) to (100)**). No need for negative numbers here. It doesn't matter that we're using the same •*"PowerIs"* variable, because it's just a temporary variable used only while a snowball is flying through the air (it changes every time). Switch the ⑦ •[**Turn Right (2) Degrees**] to a •[**Turn Left (2) Degrees**] so it spins toward their target. Lastly, in the second •[**If**], change the ⑧ *snowman* over to the *reindeer* to detect if they hit the *reindeer* with a snowball.

Now you should be able to have play alternate between the two players lobbing snowballs at each other.

Step 11: Improved Controls

Now that we can play the game switching sides and taking shots, let's make it a little easier to play by enabling keyboard controls too. Switch to the *Reindeer's* code to handle this.

Start by adding a ① •[**When** ▷ **Clicked**] and •[**Forever**] loop, since we want to always be able to enter commands for our *reindeer*. This means aiming can happen even while the *snowman* is taking a shot (outside of the player's turn). As long as we can't throw, it's no concern. Next, add an •[**If**] for four different key presses. Use the •Sensing code block •<[*Key A*] **Pressed?**> as conditions. We want one for each of the arrow keys. The ② Up and Down arrows will adjust the •*"Power"* variable up or down, respectively. The ③ Left and Right arrows will take care of the •*"Angle"*: left increasing, right decreasing. But remember, we can only have •*"Angle"* or •*"Power"* between 1 and 100. For this reason, we need to add some additional •[**If**]s to handle these limits. ④•[**If** •<•(*"Power"*) > (*100*)> **Then**] will detect if we've exceeded the *"Power"* limit, and •[**Set** [*"Power"*] **to** (*100*)] will maintain the maximum limit. Likewise, ⑤ •[**If** •<•(*"Power"*) < (*1*)> **Then**] {•[**Set to** [*"Power"*] (*1*)]} handles the minimum limit. We do the same with the ⑥ •*"Angle"*. It's a little repetitive, but this makes sure we've got nice, smooth control and stay within range.

> ### Out of Bounds
> *A common type of error in computing is an out-of-bounds number. When we're writing code, we use numbers for all kinds of things, and we can change those numbers to handle different situations. But what happens when a number is different from what we expect? Well, just about anything could happen. In some cases, it could cause a program to just stop working; in other cases, you just get unexpected or unideal results. Here we have a number for the power of a throw, and a number for the angle of the throw. In our game, we're limiting these numbers to 1 to 100. But how do we limit them? If we just allow key presses to increase or decrease the number, there's no limit, and we could go past our ideal boundaries. In some languages, you can "clamp" values to a set range, but here we have to do that manually, setting them back to the limit if they are out of range.*

Intermediate Project 3: Snowball Fight

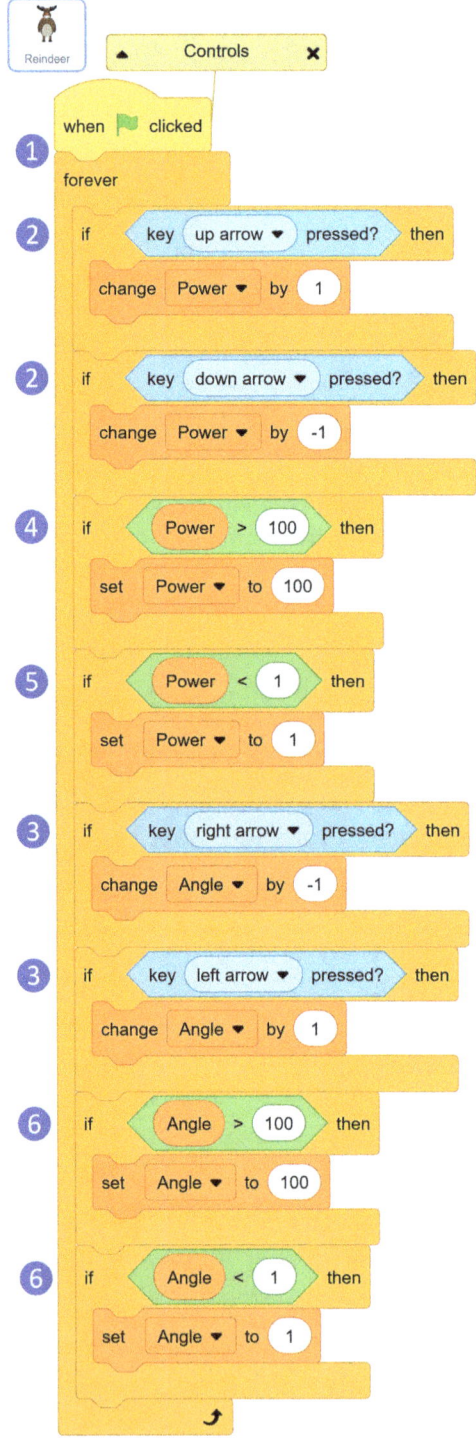

Step 12: A Palpable Hit!

Let's start adding some code to our *snowman*. Start with a ① •[When ▷ Clicked] and a ② •[Forever] loop. Inside, add an ③ •[If] and the •Sensing block •<Touching (Snowball)> set. This will trigger whenever the *snowman* is hit by the *snowball*. Next, add a new variable: •*"Points"*. This way, we can score the player on their gameplay. Keep this variable visible, right-click on the display for it and change it to **large readout**, and place it at the bottom centre of the Stage Window. Inside the •[If], add ④ a •[Change ["Points"] By (3)] and a •[Say Block ("*Ya Got Me!*") for (2) seconds] and a •[Wait (3) seconds]. We're giving more than one point for a hit here because later we'll be adding in another scoring mechanism and want to make sure there's a focus on hitting the *snowman* by valuing this achievement. The delays help prevent a hit from being counted repeatedly before the colliding clone is deleted.

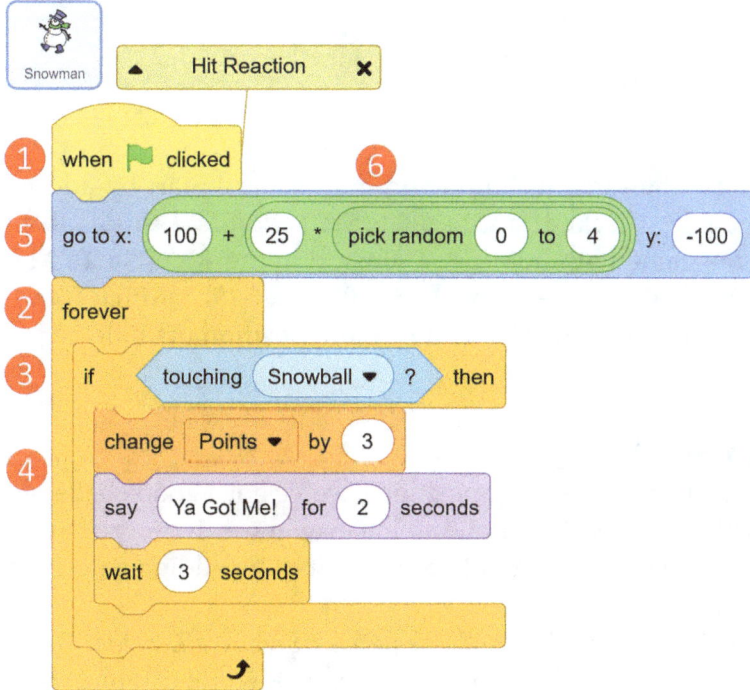

Let's make our game a little more interesting by adding a random factor. Above the •[Forever], add in a ⑤ •[Go To X: (#) Y: (#)] and set the Y to -100, but in the X, add in a random position. We'll use ⑥ three •Operators blocks to make this work: a •((0) + (0)) inside its second number, we put a •((0) * (0)), and inside its second number we put a •(Pick Random (0) to (10)). Altogether we get: •((*100*) + •((*25*) * •(**Pick Random** (*0*) **to** (*4*)))). It's a little tricky, but it'll give a range of 100, 125, 150, 175, or 200. We have a minimum value of 100, and then add 0 to 4 increments of 25. This will make sure the *snowman* is positioned at X100 to X200, so they're always on the right-hand side. You could use just •(**Pick Random** (*100*) **to** (*200*)), but we thought an incremental random would be a good technique to learn because sometimes you'll have specific uses for it (aligning to grid positions, for example).

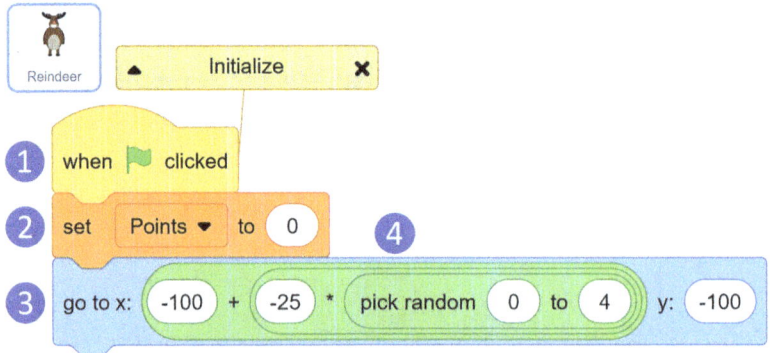

Now, use the same technique on our *reindeer*! Start by adding another ❶ •[**When** ▷ **Clicked**] and a ❷ •[**Set** [*"Points"*] **to** (*0*)] to make sure we start the game with nothing. Then add in the ❸ •[**Go To X:** (#) **Y:** (#)]. We'll have Y-100, but we'll do ❹ •((*-100*) + •((*-25*) * •(**Pick Random** (*0*) **to** (*4*)))). This will give our *reindeer* a random position of -100, -125, -150, -175, or -200.

Step 13: An Extra Challenge
With our characters changing positions, we've got a good challenge, but let's add something that will complicate life even more. If you're having a snowball fight, a snow fort is a great addition. Start by adding a new blank sprite called *SnowWall*.

In the **Costumes** tab, we'll create our *SnowWall*. Select the ❶ **Circle tool** and set the ❷ **fill colour** to white, or a white–ice blue radial gradient, with an ice blue ❸ **outline**. Start with a big circle with the bottom just touching the ❹ central target. Next, duplicate this costume, and in the next one, add a second, slightly smaller circle stacked on the first one. Repeat this process until we have four stacked balls of snow. Then, add a ❺ blank **costume** and move it to the top. Now we've got from 0 to 4 snowballs' worth of wall.

Intermediate Project 3: Snowball Fight ◆ 89

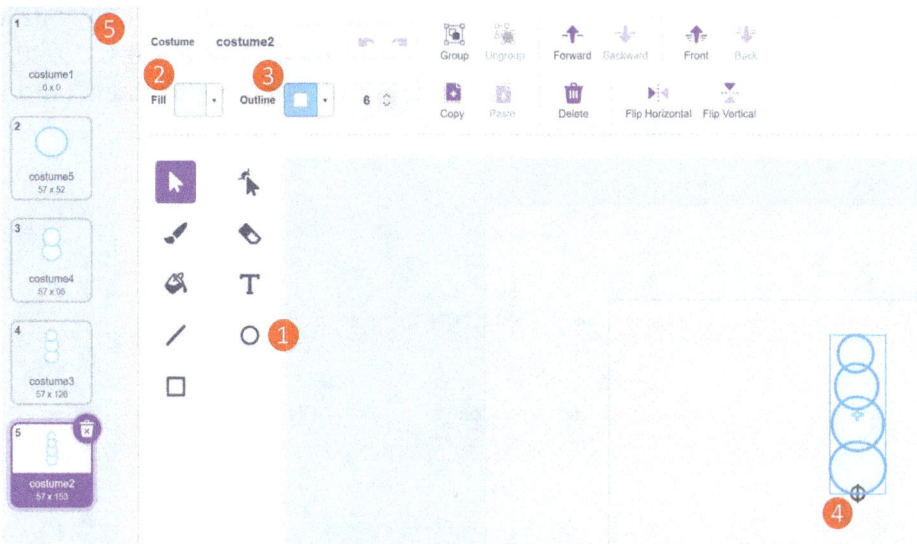

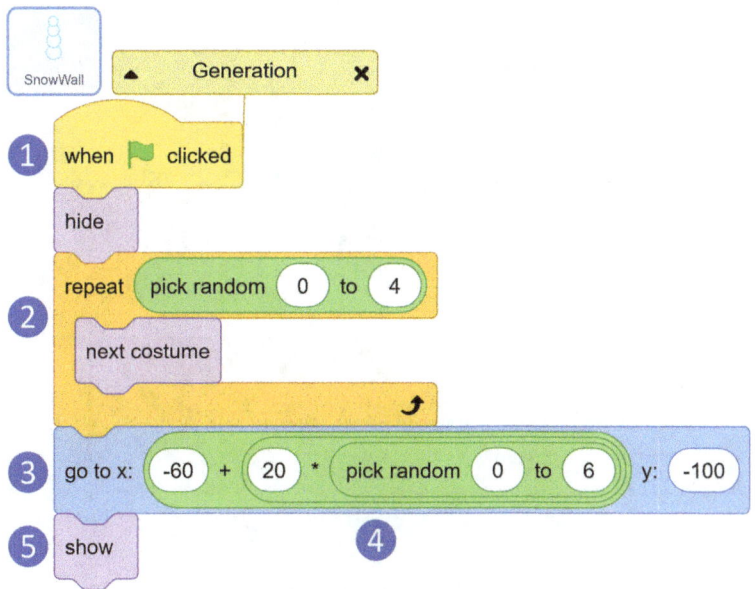

In the code, add a ① •[When ▷ Clicked]. Start by hiding our *SnowWall*, so we can get it set up correctly before showing it. We can set its costume randomly by using a ② •[Repeat •(Pick Random (0) to (4)) with a •[Next Costume] inside; this will cycle randomly between any of the five costumes (and heights). Then, use a ③ •[Go To X: (0) Y: (0)] and set it to a random position. Y will always be -100, just like our characters, maintaining a constant sense of ground level. In the X we'll use the formula ④ •((-60) + •((20) * •(Pick Random (0) to (6)))). This will give it an X position of -60, -40, -20, 0, 20, 40, or 60. So it will always be between the two characters no matter their random position. Lastly, use a ⑤ •[Show] block to make it visible.

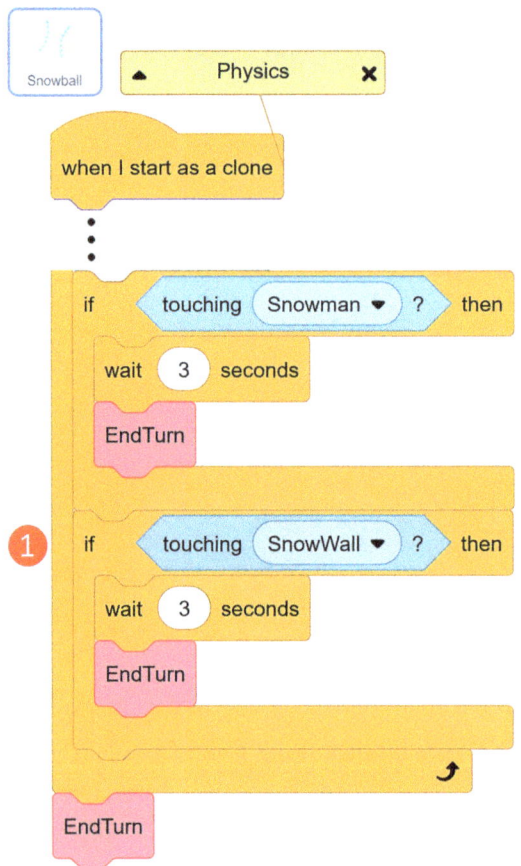

Now that we've added our *SnowWall*, let's make sure our *snowballs* will be affected by it. Use the exact same code for both *snowball* and *snowball 2*, just duplicating our ① •[If •<Touching (*Snowman/Reindeer*)?> Then] conditional with its contents and switching the sprite to *SnowWall*. This will make sure when the snowball hits the *SnowWall*, it will stop and the turn will end.

Games are fun when the challenge level is just right, not too hard, not too easy. In Book 1: Beginner, the Butterfly Catcher game is a good example of making a game that automatically adjusts to the player's capabilities.

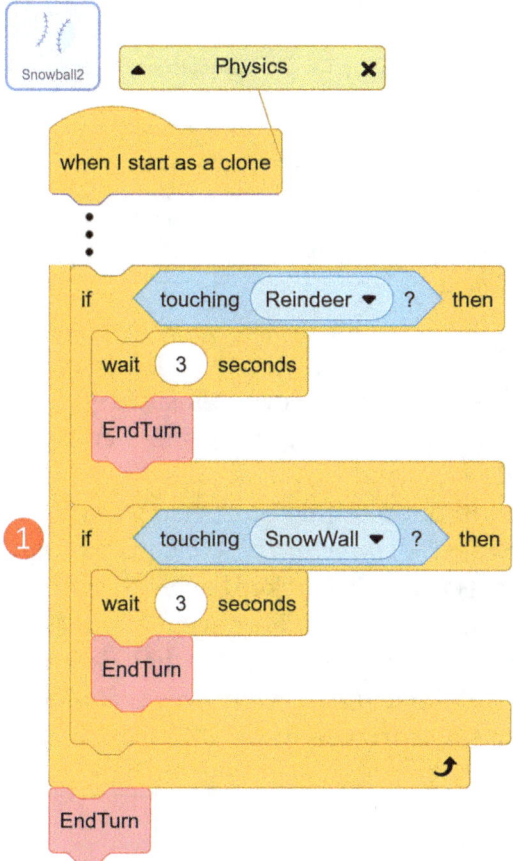

Step 14: The GameStart Event

We've got a pretty good game going, but we have to restart the game when an opponent is hit. It'd be nice to play multiple rounds without needing to restart, wouldn't it? It'd also be nice to have the positions and wall change without having to reset the game.

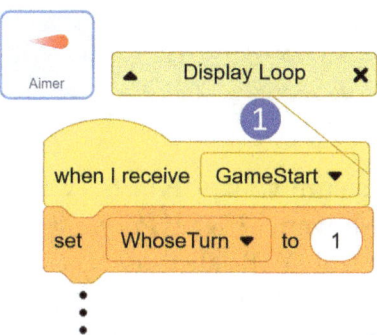

92 ◆ Intermediate Project 3: Snowball Fight

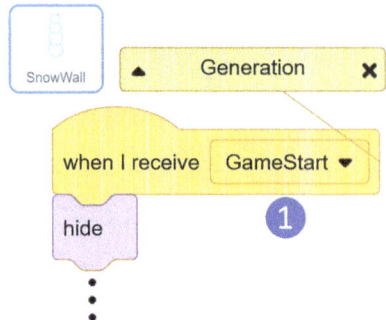

What we need is a GameStart event, instead of the game only starting once when you click the green flag. By using a custom message event, we can call it whenever we want! By having it set up correctly, we can use this to reset the match without resetting scores (so we can play forever)! Start by making a new message: •*"GameStart"*. We'll go through all our sprites and switch most of our •[**When** ▷ **Clicked**] over to •[**When I Receive ["Game-Start"]**] events. Change the •**Green Flag** event in the ❶ *aimer*, *SnowWall*, and *snowman*, and make a //*Reposition* stack for the *reindeer* by moving the •[**Go To X: (#) Y: (#)**] to a •*"GameStart"* event. This will allow them all to be reset concurrently with a single trigger event: •*"GameStart"*.

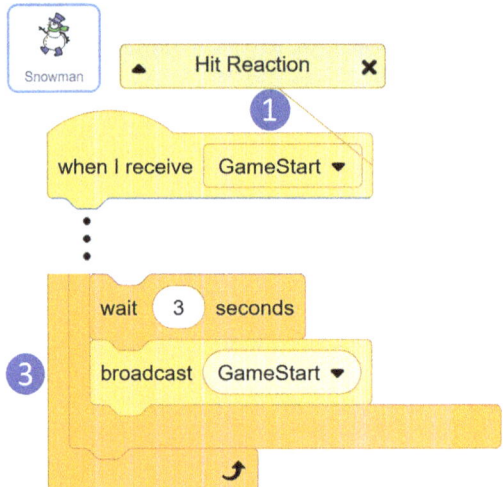

In the *reindeer*, make sure your •*"Points"* variable isn't being reset in a •*"GameStart"* event. We want •[**Set ["Points"] to (0)**] to only be under a •[**When** ▷ **Clicked**] event. With all those events switched over, you just need to call ❷ •*"GameStart"* after you •[**Set ["Points"] to (0)**] in *reindeer*. We just need one more thing to have your game play endlessly: in the *snowman*,

Intermediate Project 3: Snowball Fight ◆ 93

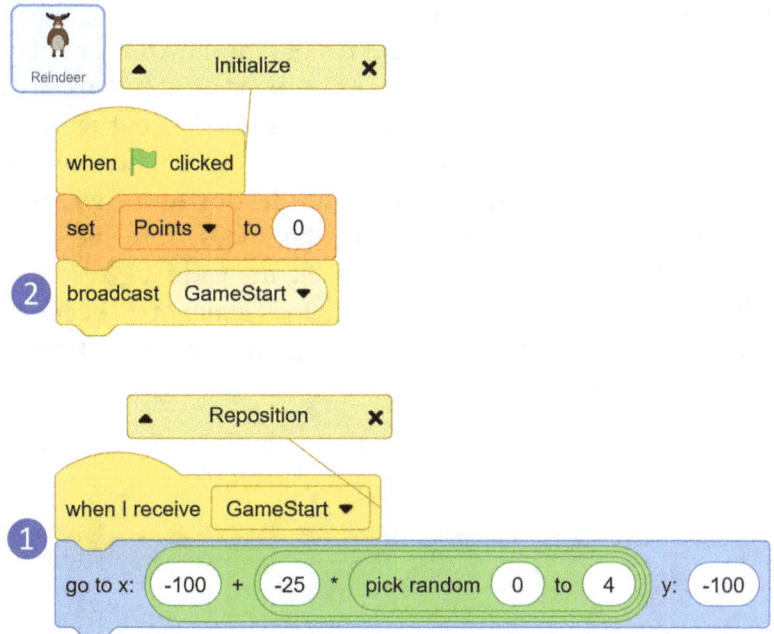

add in a ③ •[Broadcast ["*GameStart*"]] code block at the bottom of their •[If •<Touching [*Snowball*]?> Then] block. This will allow a new bout to start when the *snowman* is hit, chaining game after game when you manage to hit the *Snowman*.

There are two other complications we need to deal with to have it run smoothly, though. First, you have to get rid of any snowballs in play so they don't run their code that ends turns out of sequence and potentially mess up our order. In each snowball, add a ① •*"GameStart"* event with a •**[Delete This Clone]** code block under them. Then, make sure when the game does start again, we get some visual notification. In our *turn sign* sprite, make a new •**My Block:** •*"ShowTurn"*. Put all the codes from the •*"NewTurn"* stack into the ① •**[Define *"ShowTurn"*]** stack, then add a ② •*["ShowTurn"]* block to the *//New Round* stack and add a ③ •*"GameStart"* event and add a •*["ShowTurn"]* block to that. This way, we get notice whose turn it is if we start a new bout or a new game.

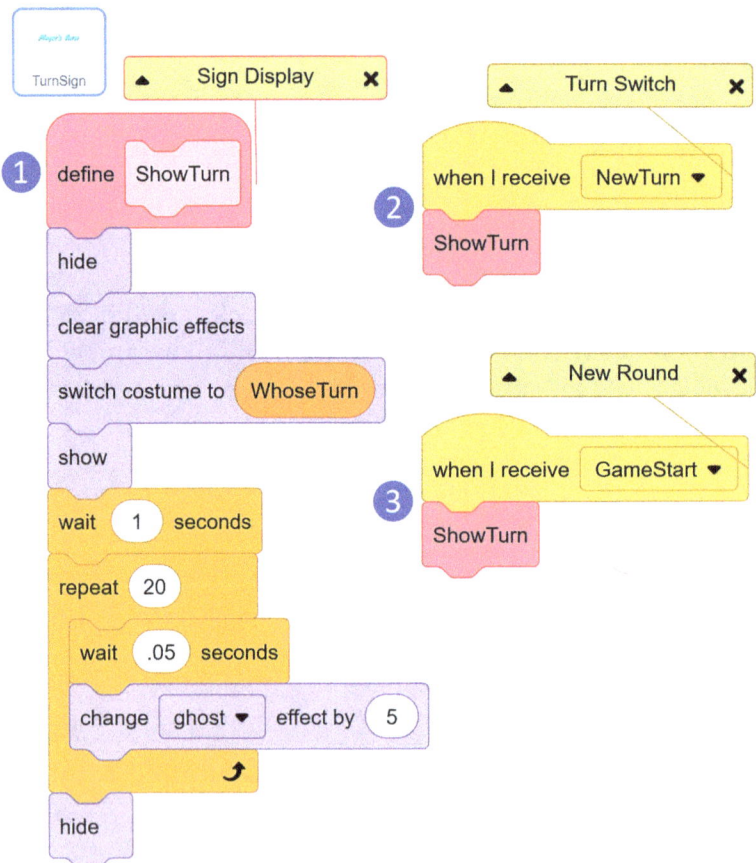

Step 15: Wind Effects

Now that we can see how fun our randomization is, let's add one more classic feature to play with: wind. Start with our *snowball*. We'll need a new variable, called •*"Wind"*, displayed at the bottom centre of our Stage Window. Add into our •*"GameStart"* event a ① •**[Set *["Wind"]* to** •**(Random (-10)**

to (10)]. With it, every game will have a random •"*Wind*" factor. Next, add a
②•**[Change X By (#)]** code block by our gravity •**[Change Y by (#)]** code
block. Here, instead of a set value, we'll use ③ •(•("*Wind*")/(8)). This will
allow our •"*Wind*" factor to slowly move our snowball along in its direction.
As a reminder, because this is happening every step, it can be a very small
number and still have a big effect.

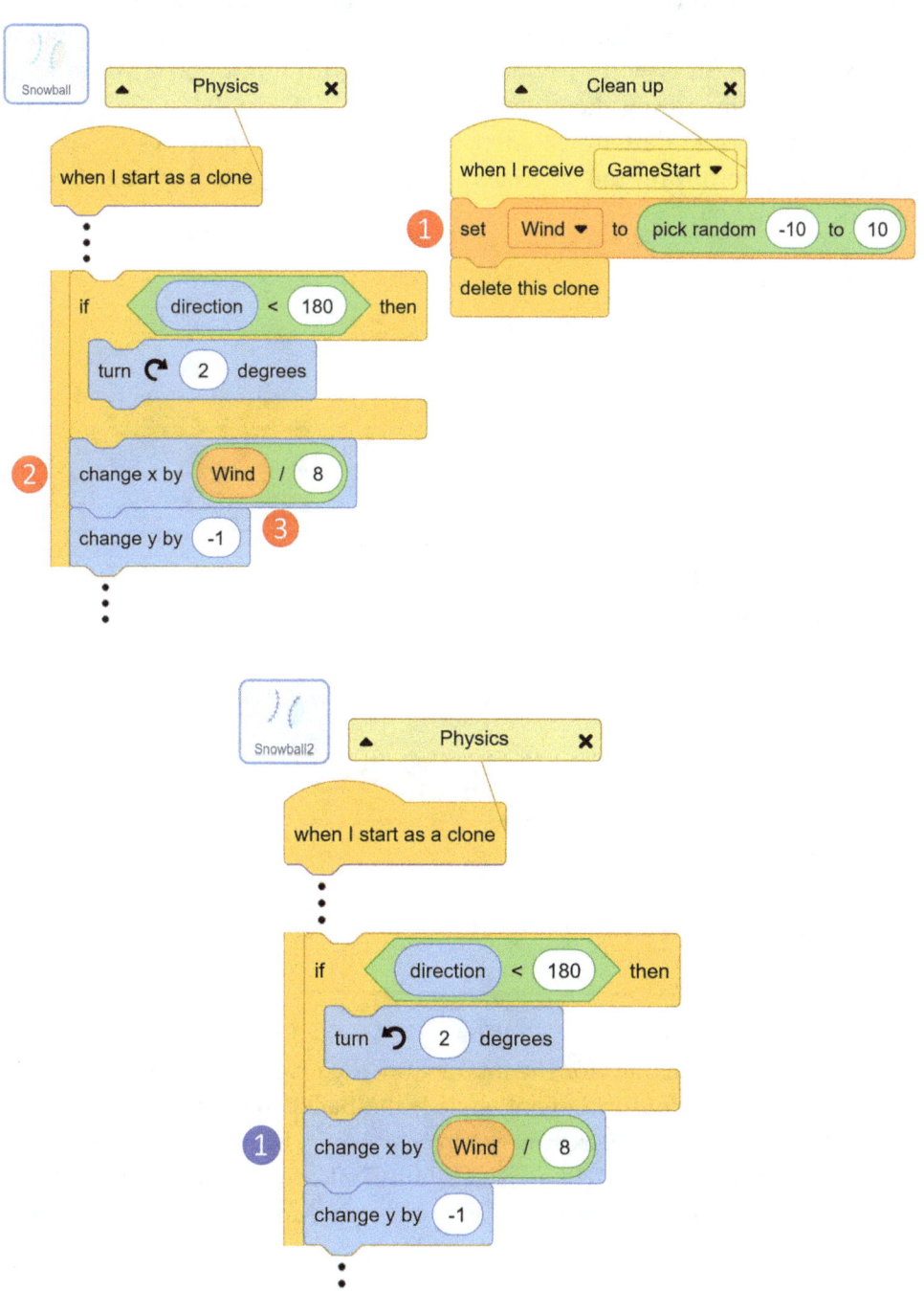

Lastly, add the same ①•**Change X** combo to our *snowball 2* effect so the *snowman*'s throws are also affected by the wind.

Step 16: Bonus Stars

We've got a pretty good snowball fight going on, but we can still make it even more interesting with a little tweak. Even if we can get pretty good hitting a single target reliably, what if we had multiple targets to keep track of? Let's add the *star* sprite from the Sprite Library and turn it into a bonus point-scoring feature! Stars will provide the player with an opportunity to score extra *"points"* by hitting them with their *snowball*. Since they won't stop the *snowball*, players can attempt to arc their shots in specific ways to hit stars on their way to hitting the *snowman*, presenting an extra challenge for our advanced players!

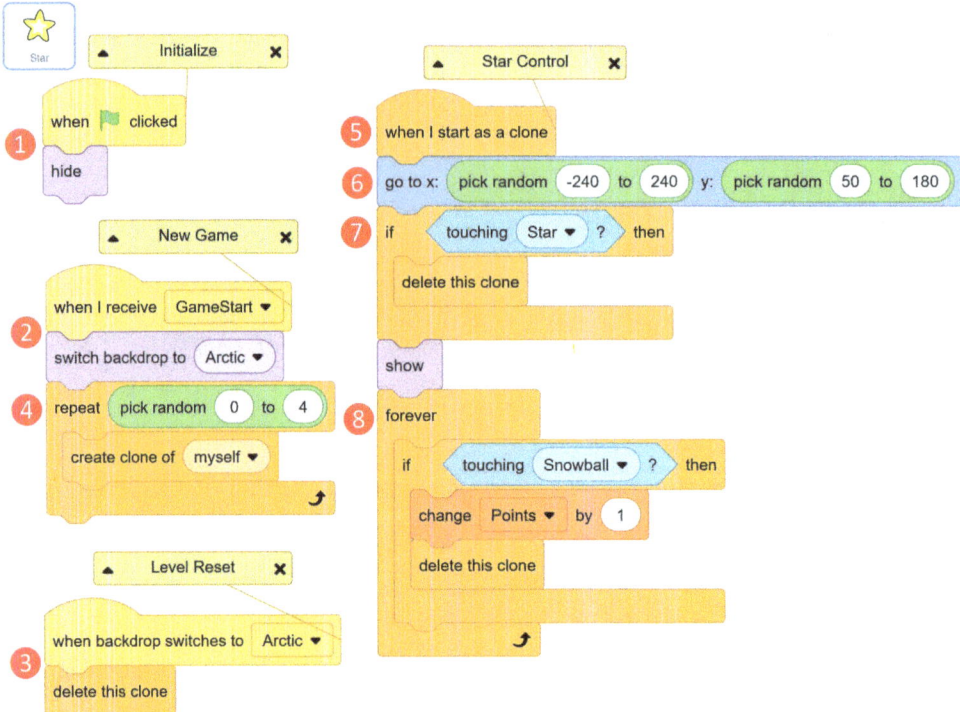

In the *star* we'll need a ①•**[When ▷ Clicked]** to •**[Hide]** the original object. While using clones for the interaction, the original will be hidden.

In our ② *"GameStart"* event, switch the **backdrop to Arctic**. This may seem strange, since the backdrop is already Arctic, but we're just using this to trigger another event. Add a ③•**[When Backdrop Switches To [*Arctic*]]** (which runs even if the backdrop was previously Arctic), and add a •**[Delete**

Intermediate Project 3: Snowball Fight ◆ 97

This Clone]. The reason we do this is to clean up any clones from the previous bout; otherwise, we'd end up stacking them endlessly.

Going back to our *"GameStart"*, add a ④ •**[Repeat** •**(Pick Random (*0*) to (*4*))]** with a •**[Create Clone of [*Myself*]]** inside it. This will ensure 0 to 4 stars are generated on each bout.

Next, we'll program the ⑤ •**[When I Start As A Clone]** event. It needs a randomized position, but we can't just use the normal •**[Go To [*Random Position*]]**, because the snowballs can never go below Y: -100. So here we'll use a ⑥ •**[Go To X:** •**(Pick Random (*-240*) to (*240*)) Y:** •**(Pick Random (*50*) to (*180*))]**. This ensures the stars will be up in the air, where the snowball can hit them.

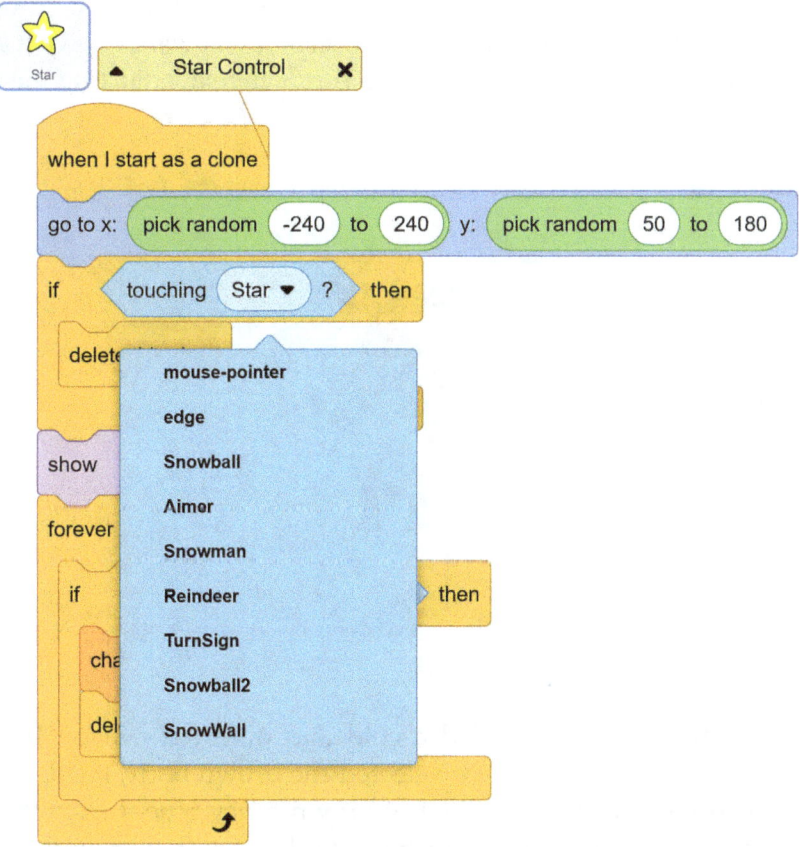

Under that ⑦, we'll test if the *star* is touched by another *star* and, if it is, •**[Delete this Clone]**. You'll need to actually make this code in another sprite (like the *reindeer* or *aimer* – it doesn't matter which). Select •**<Touching (*Star*)?>** and then copy the stack to the *star* because Scratch doesn't let you test if a sprite is touching itself normally (if you try to make the code in the *star*, the •<**Touching (*Sprite*)**> code block won't list *star* as an option, so we have to use a different sprite to create it). We want the stars to be spread out, since overlapping stars are too likely to be both hit at the same time. We want to reduce those cheap •"*Points*". Once in position, •[**Show**] the star and then have a ⑧ •[**Forever**] loop. In the loop, test for collisions with the *snowball*. If it's a hit, the player will earn a point and the clone will be deleted. Voilà, our secondary scoring system is here to add an optional challenge to players!

Step 17: Defeat

As in any game, we should also be able to tell when the player loses; otherwise, our high score would just be a matter of time, not skill or luck. Let's add a blank sprite and call it *GameOver* and position it at X: 0, Y: 0. In the **Costumes** tab, draw out the Game Over screen. If you want to give it a snow-covered look, use the technique from the *TurnSign* text, adjusting the colours and position of the text.

Underneath the text, I added a cold and defeated *reindeer* by changing the mouth into a frown, then selected the brown body, head, and arms and changed their **fill** to a gradient of brown to icy blue. I used the **Circle tool** and then heavily used the **Reshape tool** to add new points and drag them all over to make some icicles hanging from the antlers and made them white–icy blue gradient-filled. Try switching points between the **curved** and **pointed** settings to see how that affects the shape. Lastly, I added a couple of icicles hanging from the chin. This is definitely one frozen, slush-soaked *reindeer* ready to call it quits and hit the hot chocolate. We can now switch to its code.

Intermediate Project 3: Snowball Fight ◆ 99

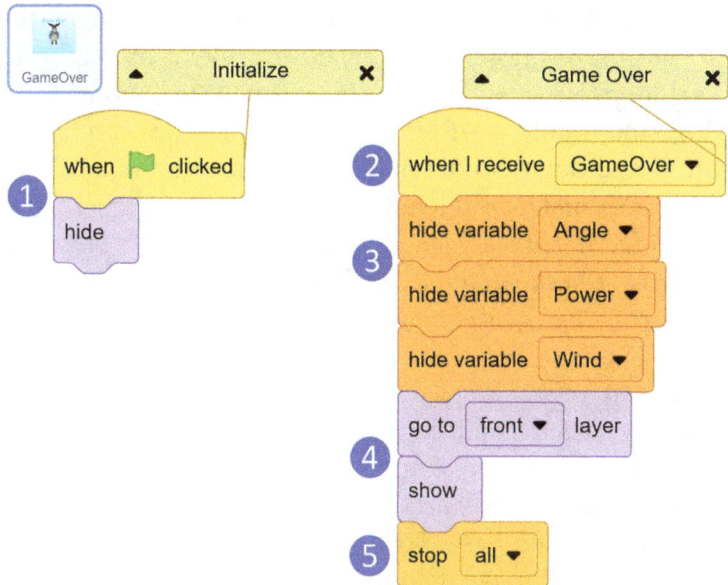

Start with a ① •[When ▷ Clicked] and a •[Hide] since we definitely don't want our game starting over already. Then, add a ② •*"GameOver"* event. We'll ③ hide our variables being displayed, except for the •*"Points"*, so players can see how they did. We'll add a ④ •[Go To [*Front*] Layer] and •[Show] to display the Game Over screen overtop of everything, and add a ⑤ •[Stop [*All*]] to end the game.

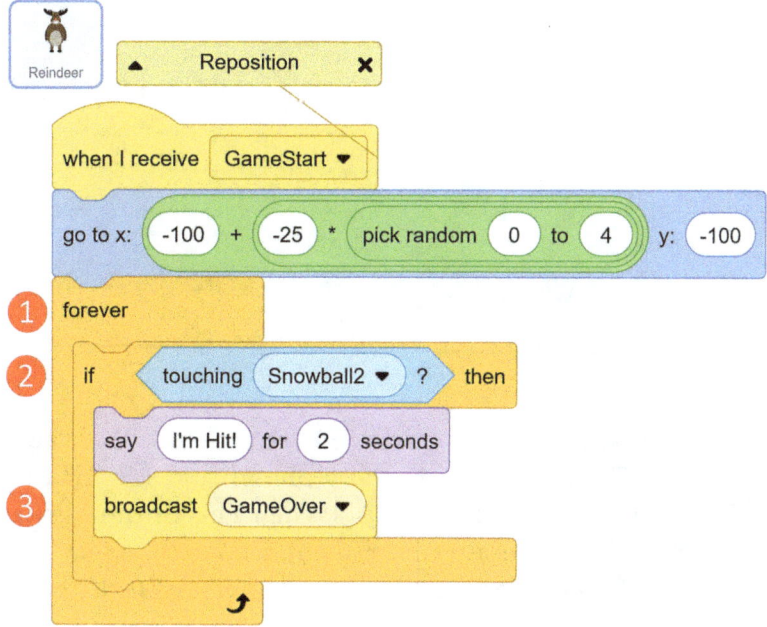

Lastly, we just need to call the •*"GameOver"* event. For that, head to our *reindeer*. In our positioning •*"GameStart"* event at the bottom, add a ❶ •**[Forever]** loop. Here, add a ❷ •**[If <Touching [*Snowball 2*]?> Then]** conditional. Inside it, add a •**[Say "I'm Hit!" for (2) Seconds]** to give a little exclamation and delay, so the player can perceive the hit. Then add in a ❸ •**[Broadcast ["*GameOver*"]]** to end the game.

There's a lot of events in this game that could use some extra pizzazz to take them to the next level. You can check out the Scrolling Shooter project in Book 3: Advanced for some special techniques on adding graphical flair and design to your projects!

Step 18: Start Screen

Very last thing we need is add a *Start screen* to our game welcoming our players. Add a new blank sprite and name it *Start Screen*, and position it at X: 0, Y:0. In the **Costume** tab, start designing a title image. Add a pure white rectangle background as a base. Then, add a rectangle covering the top half or less of the screen and make it an icy blue to white gradient with white on the bottom. We'll also make another rectangle covering the bottom half or less and make it icy blue to white gradient with white on the top. This creates a cool white beam effect across the middle of the image. Using the text technique we've used earlier, add a title with two separate words, "Snowball" and "Fight", enabling separate positioning and centring of each word. Lastly, add a little text line telling the user, "Click to Begin". Next, we can code it. If you want, you can add drop shadows like in Step 8: The Turn Sign.

We'll need a ❶ •**[When ⚑ Clicked] Event** to start. We want the *Start screen* to ❷ •**[Go To [*Front*] Layer]** and •**[Show]**, and ❸ •**[Hide Variable [*variable*]]** all the variables. Add in a ❹ •**[When This Sprite Clicked]** event. Here, we'll ❺ •**[Show Variable [*variable*]]** for our •*"Angle"*, •*"Power"*, •*"Points"*, and •*"Wind"*. Then ❻ broadcast •*"GameStart"*, •**[Hide]**, and broadcast *"NewTurn"*. As the very last step, ① remove the [**Broadcast** ["*GameStart*"]] from the *reindeer*, and because we switched everything else over to the *"GameStart"* events already, our game should be ready to go!

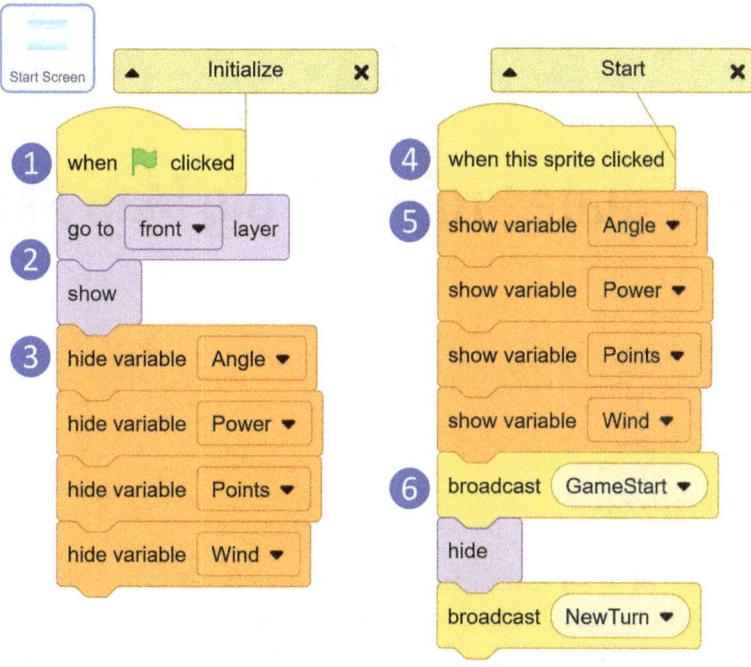

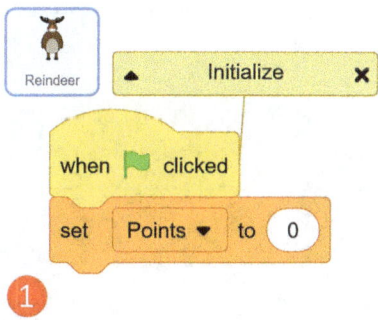

For working copies of this and every project in the book series, visit www.massivelearning.net *for direct links to Scratch projects, and to see our other projects and resources for coding education!*

8

Intermediate Project 4: Big Map Racing

What This Project Is

This is a single-player racing game with a huge Scratch limit-breaking map. The player will control a race car while speeding around the giant map to get the best possible time score. They'll need to watch their driving, as off-roading will slow them down a lot, and crashing into walls will require recovery, resulting in losing precious time. We'll learn some clever techniques over about an hour and a half to complete the project.

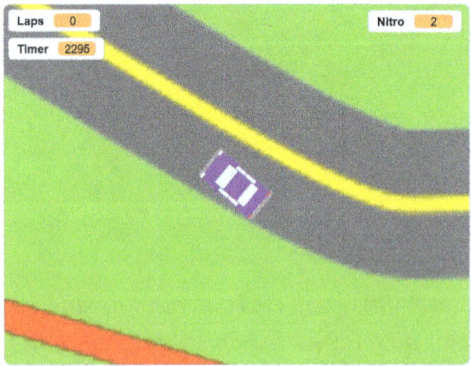

Intermediate Project 4: Big Map Racing ◆ 103

What We're Learning with It

In this project, we're going to bump up against some of the safety features and limitations of Scratch while exploring methods to get around them. We'll be pushing the limits to both enhance our gameplay experience and to learn and use some more complex coding methods. Doing so, we'll explore a lot of classic game mechanics and techniques, including movement systems, object referencing, player states, timers, power-ups, and menu settings. This project has a lot of learning outcomes, and these techniques will help flesh out our intermediate skills and get us set up for our advanced projects.

> **Conditional Movement.** Adapt movement rates based on conditions to allow for speed penalties when off-roading.
> **Colour Collisions.** We'll use colour collisions to create a terrain system using an object that functions like a backdrop.
> **Waypoint Testing.** An anti-cheating feature will be put in place to make sure players complete full loops of the track for it to count as a lap.
> **Size and Position Limitations.** We'll learn the limits of size and position in Scratch that prevent you from working with very large or small objects and keep everything visible on screen.
> **Scale Breaking.** A technique to get around size limitations, enabling the creation of really big maps.
> **Referencing Objects.** We'll learn how we can use the properties of one object in the code of another, letting us dynamically react to objects for more than just colliding.
> **Player States/State Machines.** Used to determine movement and crashing, applying our game state concept to single objects.
> **Costumes as States.** Costumes will be used to determine states, showing how we can use object properties, not just Variables, to determine states.
> **Timers.** A basic timer to track how long the player takes to complete a lap.
> **Power-ups.** We'll add a nitro boost feature to the game, allowing the player to gain extra speed for a limited time.
> **Use Costs.** Additionally, we'll have a count limit to the nitro boost as a control method for usage. Great for ammo, supplies, money, fuel, or energy cost systems.
> **Menu Settings.** Lastly, we'll create a more complex menu with options affecting play.

Building It

Step 0: Create Your New Project
Make sure you're logged in to Scratch, then click Create to begin a new project! Since we won't be using it, we can delete the Scratch Cat sprite by clicking on the trash bin on that sprite's thumbnail in the Sprite Listing.

Step 1: Drawing a Racetrack
Our first task will be to make the *racetrack* sprite. Starting from a blank sprite, name it "Racetrack" and position it at X: 0, Y: 0. We can ❶ rename its first costume *"Track1"*. Create an ❷ oversized rectangle of grass green or mud brown (you'll want a unique colour to convey the off-road condition). Then, select the ❸ Circle tool and set the ❹ **fill** to transparent and the **outline** to 30pt wide and a grey colour. Draw a large circle. Next, use the ❺ **Reshape tool** to add points and stretch and reshape it to make a circuit of road for the race to take place on. You may want to explore with the curve angle controls – the two little lines and dots that extend out of each **curved** point. Moving the dots on these changes how the curve is calculated, making long or short curves and changing the twist of the curve. It's a very powerful tool but takes a lot of practice. A good racetrack will have a mix of straightaways, curves, and hairpins. If you want a centre line for the track, you can just copy the track, move it overtop of itself, change its ❻ outline colour to yellow or white, and reduce its outline thickness to a very small number. By doing so, you'll get the exact same shape, and with a little adjusting, you can get the line perfectly in the middle for the whole track.

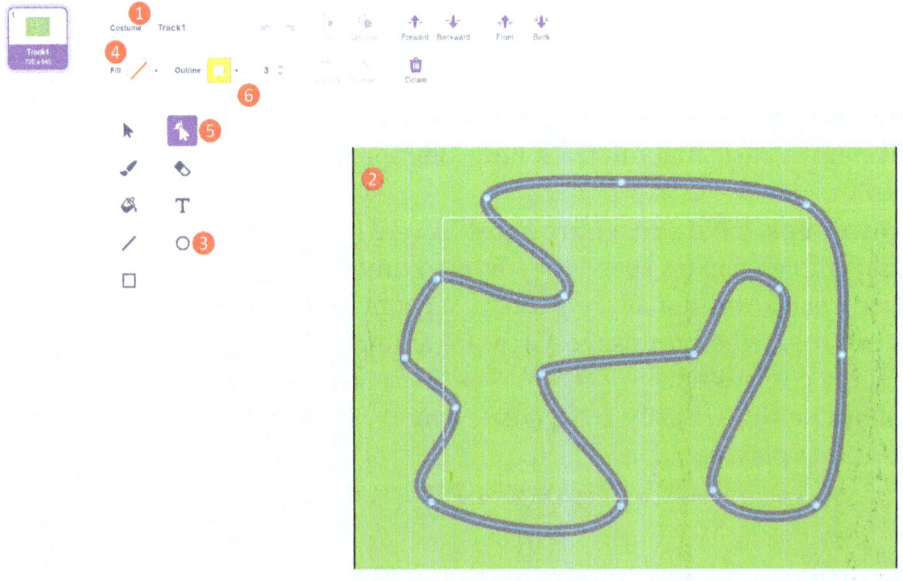

Now, let's try to resize our track by making it 10× larger! Type 1,000 in the **size** property and hit Enter. Didn't quite work, did it? This is one of the protection protocols in Scratch. To make sure no object ends up too large or small to recognize, there's a limit on the **size** %. However, there's a way around this. First, add a second blank **costume** to our track and rearrange it to be the first costume, since we'll need it to start things. Next, add a small rectangle, enough to cover the centre target, but small enough that if it grows 10×, it will still fit in the Stage Window. The colour and outline don't matter. This is a lot smaller than our other costume, isn't it? You can name the costumes so you'll be able to keep them straight when coding.

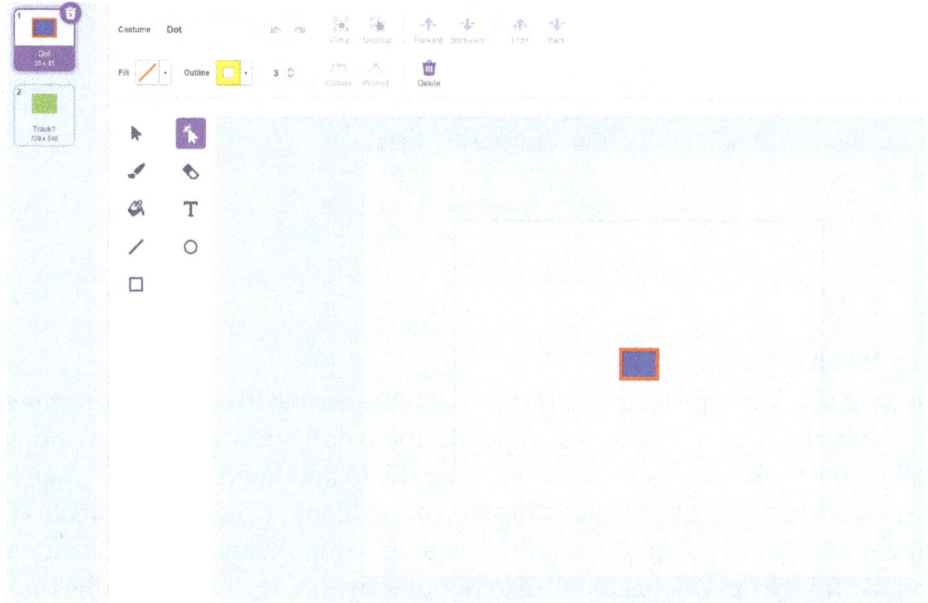

Now for some *racetrack* basic starter code. We'll be adding a menu to this project, so let's start things with a ❶ •[**When I Receive [***"GameStart"***]**] event so there's no need to come back and change it later. Now, we're going to make use of our strange alternate costume. We need four •Looks code blocks, so you might as well grab them at the same time while you're in the •Looks category. Let's set the map to ❷ •[**Go To [***Back***] Layer**], since we always want it behind everything else. But how do we deal with the size limitation? The trick is that the size limit test is based on the current costume! So ❸ switch costume to the small rectangle, then •[**Set [Size to (***1000***)%]**], and then switch costume to the racetrack. Ta-da! That is one big track. Make sure you set the track to ❹ •[**Go To X:(***0***) Y:(***0***)**] and to •[**Set Rotation Style [***Don't Rotate***]**].

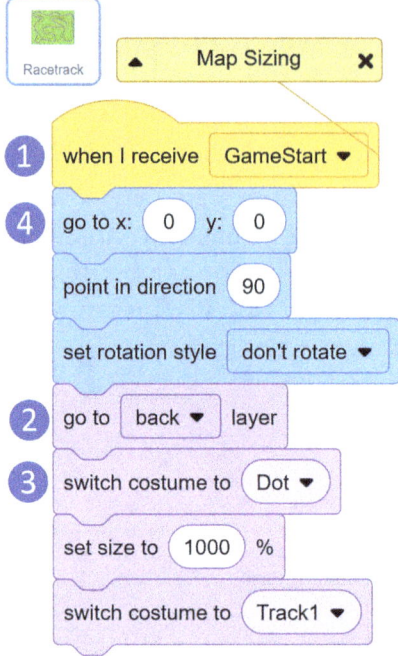

Step 2: Drawing a Car

Create a blank *car* sprite. If you're not much of an artist like myself, a rectangle for the body, with a couple rectangle bumpers and a few rectangle windows will do the trick. Make sure the front windshield and hood are bigger than the rear, or add headlights/brake lights so you can convey the facing direction of the car easier. But by all means, feel free to let your creativity loose, or try the Upload Image option. Just be sure to remove any background from the image so it's just the car itself.

Onto some basic starter code for the *car*, we need a ① •[When Clicked] to trigger our start, at least until we make our menu later. To size your car appropriately, I used •[Set Size to (30)%], but it will vary depending on how big you drew the car. Next ②, have the *car* •[Go To X:(0) Y:(0)]. You can •[Set Rotation Style [*All Around*]], although that is the default setting, to see the difference between the *car* and the *racetrack* as examples of those two settings. Add ③ in a 1 second wait so users can get their bearings before we •[Broadcast ["*GameStart*"]].

If you'd like a bit more explanation of the art tools, you can check out Book 1: Beginner and the Firework Display project as a good place to begin!

Step 3: Turning the Car

Now that we've got our car and map in place, let's figure out movement. In the *car's* code, let's add a //*Steering* stack for some controls. When we

Intermediate Project 4: Big Map Racing ◆ 107

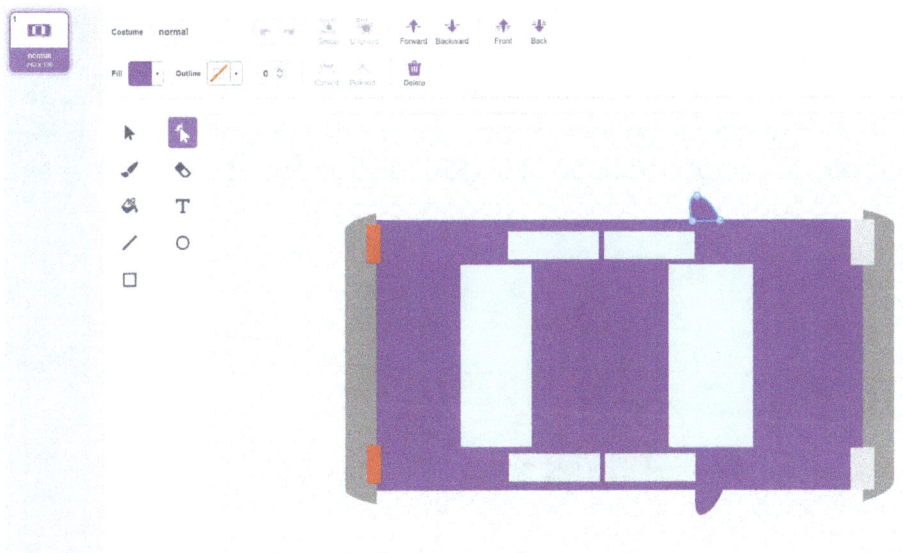

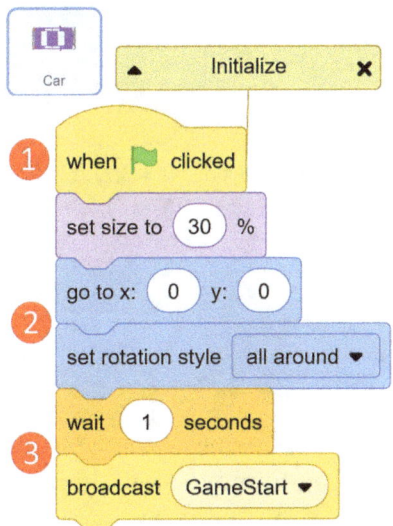

① *"GameStart"*, add a ◦[Forever] loop to handle the car's movement. Also, add a ② ◦[Go To [*Front*] Layer] to make sure our *car* will always be on top. Then, use a conditional to test for movement. In many previous projects in this series, we've used ◦Key Press events, but they tend to be a little slow/choppy for a game like this. Instead, we'll use a method that gives smooth, responsive controls perfect for driving. An ③ ◦[If ◦<Key [*Right Arrow*] Pressed?> Then] will allow our project to detect controls every step of the game and respond immediately. Nested in this ◦[If], we'll put a ◦[Turn Clockwise (5) Degrees].

Then, duplicate the •[If] code blocks we've just worked on and switch its parameters for the left key with a counterclockwise turn. Now we've got our car turning! While we're here, we can add a hidden (unchecked) •Variable needed for the next technique called •"Speed". In the •"GameStart" event above the •[Forever], add a ④ •[Set [Speed] to (10)] block and we'll be ready to move.

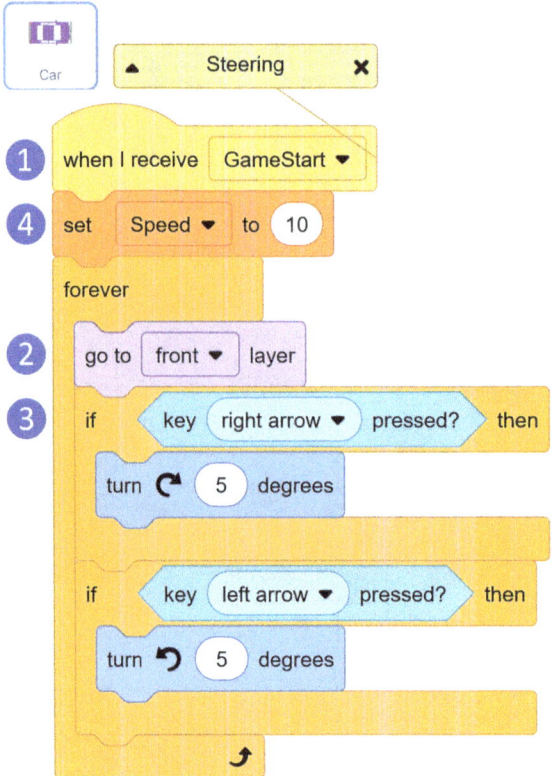

Speaking of movement, here's one of the weird things about this project: technically, our car isn't going to move at all. Instead, it's going to stay right there in the middle of our screen. To give the sense of movement, we're going to move the universe around it! To accomplish such a feat, our *racetrack* needs to move to shift in proportion to our controls. Since our map is so big, we couldn't drive all over it without moving it to be in view anyways. Go to our *racetrack*. At the bottom of its •"GameStart" event, add a ① •[Forever] loop. Add a •[Go To [Back] Layer] to make sure it always stays at the bottom. Next, we need it to turn to respond to the *car*.

Add in a ② •[Point in Direction (#)]. But what direction should it point to? Since the *car* is what's getting steered, we can refer to its properties. In •Sensing, you'll find a •([Backdrop #] of [Stage]) code block. It might

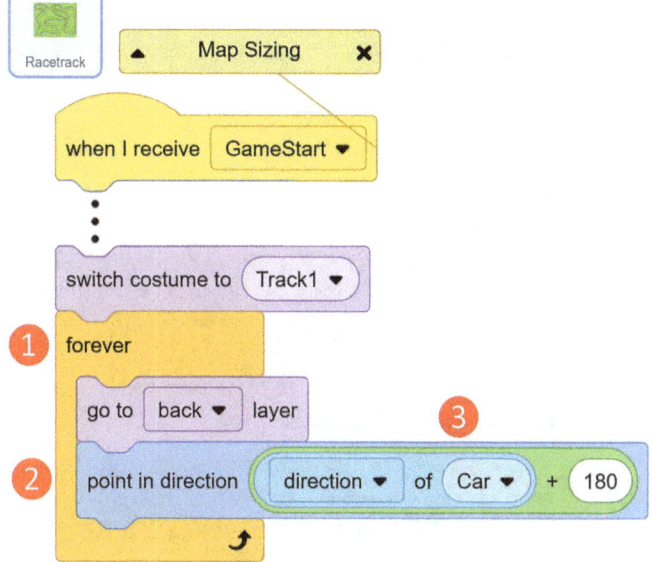

not look like what we need, but this code block is deceptively powerful and handy. By clicking "Stage" (the second dropdown item on that Sensing code block), you'll notice it can be switched to *car*. After that, click on the first dropdown item and select "Direction". This block lets you use the value of properties from other objects; when you use it, just remember to change the object first and you'll be able to see the properties you can refer to. Lastly, our *car* is facing a particular direction, but if we're moving our map, it has to go in the opposite direction. Add in a ●((*0*)+(*0*)) (Addition) code block so that we ③ ●[Point In Direction (●(●([*Direction*] of [*Car*]))+(*180*))]. Our *racetrack* now faces the opposite direction of the *car*.

Step 4: Moving the Car/Map

Because we are using an oversized map and our car is going to remain in the centre of the screen throughout the game, we need to move our map in order to "move" our *car*. While our car handles turning with the left/right arrows, we'll use the *racetrack* object to handle forward/backward motion using the up/down keys.

In the *racetrack's* ●[Forever] loop, add in an ① ●[If ●<Key [*Up Arrow*] Pressed?> Then] to generate the forward movement. Inside the ●[If], put a ② ●[Move ("*Speed*") Steps] combo. But while moving the map, what if we get to the edge of it? Inside this ●[If], we'll apply another test to prevent going off-map. ③ ●[If ●<Not ●<Touching [*Car*]?>> Then] will verify if the map has moved beyond what it should have, then in this ●[If] we need to undo the

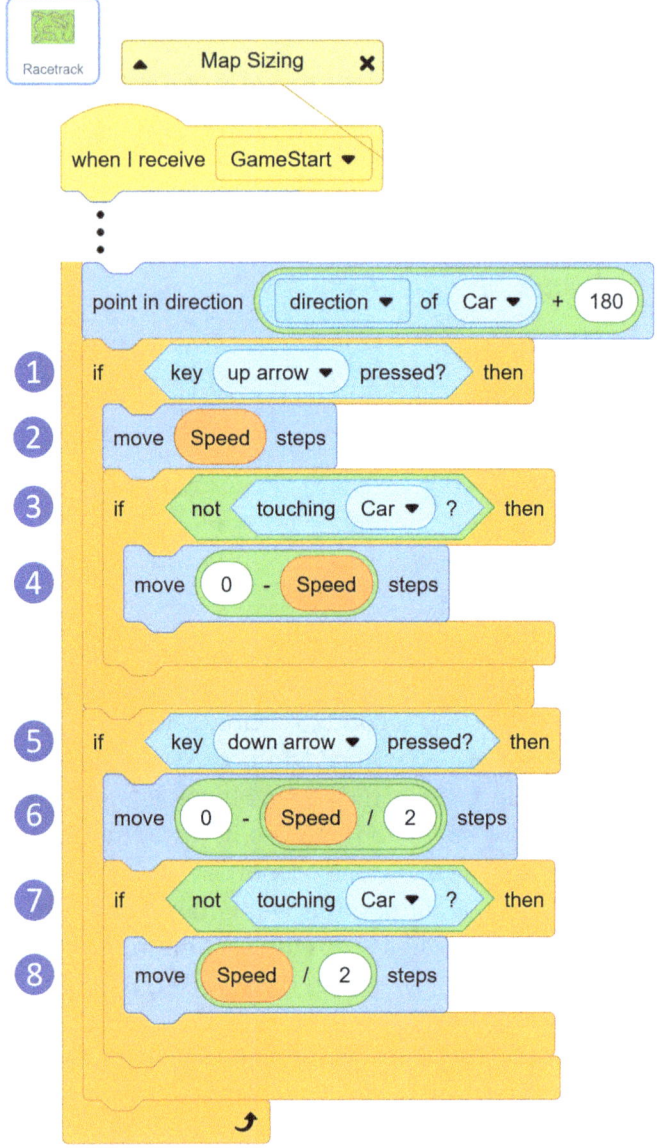

movement it just took by using a ④ •[Move •((0)-•("*Speed*")) Steps] combo – the negative of our speed will reverse the position change of our last move.

Using what we've learned to move forward, apply the same principles to moving backward, except we'll make it work at half the speed. To do so, add a ⑤ •[If •<Key [*Down Arrow*] Pressed?> Then]. Then, have a ⑥ [Move •((0)-•(•("*Speed*")/(2))) Steps] block to move negative steps (backwards) and half speed •(•("*Speed*")/(2)). Again, we need to test if we've moved off the map with a ⑦ •[If •<Not •<Touching [*Car*]>?> Then], and inside it add a ⑧ [Move •(•("*Speed*")/(2)) Steps] block to reverse the movement if that's the case.

Step 5: Off-Roading

A big part of racing is trying to find the optimum route through the course with the tightest turns. But if you can cut corners, you're bound to do better than sticking to the road. To make things a little more realistic, we can add a penalty for off-roading. If a car drives off the road, it can have its speed decreased. To do this, go to the *car* and its •[Forever] loop.

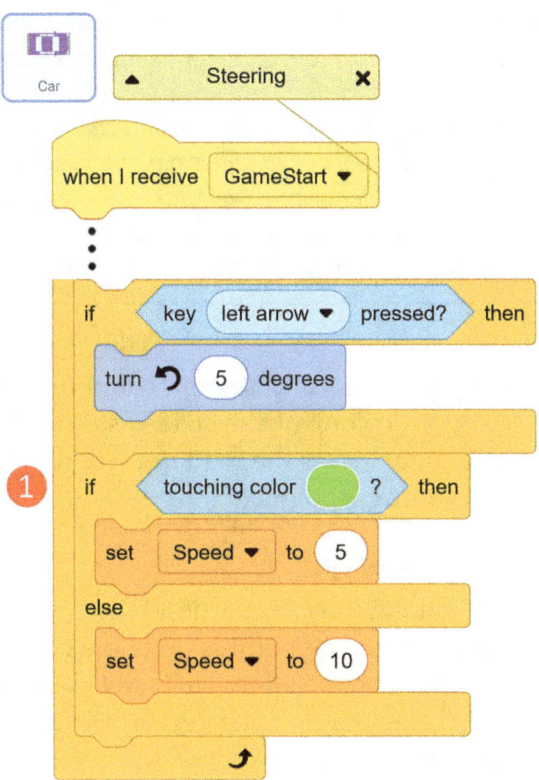

Add an ① •[If <*condition*> Then {} Else{}] block, then in •Sensing, we can use it to test if we're •<Touching (*Colour*)?> and click on the colour dot to select the specific colour of the grass on the racetrack. If the *car* is touching grass, •[Set ["*Speed*"] to (5)] to halve its speed. In the •Else, {•[Set ["*Speed*"] to (10)]}. Now, all players will want to make sure they stay off the grass.

> **Colour Detection**
>
> *Since we'll be using a lot of colour collision detection in this project, let's take a moment to talk about it. Colour collisions work by detecting a specific colour you choose in the code block. This colour is very precise! It will only detect that exact colour (similar doesn't count)! It will need to match the exact same colour, saturation, and brightness ratings. If a sprite is at less than 100% size, the rescaling of the sprite's graphics may have blended the colour of its costume graphics into something unexpected in the Stage Window. The edges of shapes, especially those resized, can be "anti-aliased," which is a computer graphics term to mean a type of blending that makes things look smoother and more natural. Have you noticed how some curved or diagonal lines in some digital graphics look jagged, while others look smooth? The difference is anti-aliasing, which basically blends a pixel's colour with its neighbours to create more gradual changes. This can be great for making smooth graphics, but it means our exact colour for collisions might not be where we thought it was! If you're using colour collisions, make sure any shape with the colour you want to detect is big enough that it will have pixels that aren't anti-aliased into a blend, making it hard to detect! Later in this project, if you're laps aren't being counted right, I can almost guarantee it's because of anti-aliased colours in the game not matching the colours in your code.*

If you want a less-penalizing system, you could use **<Not <Touching (*Colour*)?>>** and select the colour of the road instead. This will only slow down the *car* if it completely leaves the road but will require some consideration in how we handle our waypoints in the next step.

Step 6: Waypoints and Lap Counting

Our race will be run by timing three laps around the track. To achieve this lap counting, we're going to need a waypoint system. If we instead simply added a finish line, players could drive in a tight circle to touch it in rapid sequence to score laps. So how do we make sure players actually drive around the track for it to count as a full lap? If we have hairpin turns, how can we make sure they don't just cut across? We can solve these issues with a surprisingly simple system in our *car's* code.

Start by adding a ① **"GameStart"** event for our *//Lap Counter* stack. We'll add a visible (checked) variable **"Laps"** to track our current lap. Start our **"GameStart"** stack with a ② **[Set ["Laps"] to (0)]**. Use the ③ **[Repeat Until <("Laps")=(3)>]** loop. This will let us run a loop until we've run enough laps. Inside it, we'll have a series of ④ **[Wait Until <Touching (Colour)?>]** blocks. To make this work, we need to adjust our track to have lines of colour

Intermediate Project 4: Big Map Racing ◆ 113

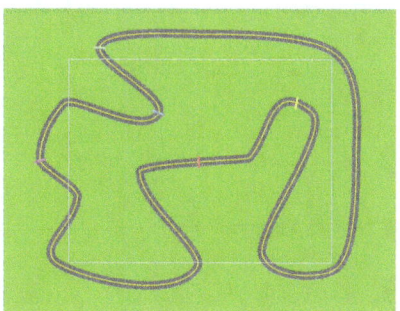

positioned across the road (on top of it) to ensure the player runs the course as intended to touch all of them in order. Make sure the lines are thinner than the *car's* length, so the car is always able to touch the grey road while crossing them, avoiding our off-road system from *Step 5*; an outline width of 6 tends to work well. You can add as many •**[Wait Until** •**<Touching (Colour)?>]** as you want, but keep in mind you'll need the same number of coloured lines and in the same sequence in any and all racetracks you create. If you want to make life easier for yourself, always have the start/finish of the *racetrack* be at X: (0), Y: (0). Remember that you can temporarily shrink your *racetrack's* size property to get everything back on-screen and make colour picking easier, or to run a test race much faster.

The code waits until the touching colour has occurred before proceeding, so each colour will need to be touched in order to proceed, starting from the first. You should use your first colour as your start/finish line, and duplicate that specific •**[Wait Until** •**<Touching (Colour)?>]** to also be at the bottom of your list, making same colour start and finish the sequence. Under all •**[Wait Until** •**<Touching (Colour)?>]** combos, but still inside the •**[Repeat Until <condition>]**, include a ⑤ •**[Change [*"Laps"*] by (1)]**. This way, when the *car* successfully touched all the colours in sequence, a lap will be finished. After finished all three laps, it will proceed to run the code after the •**[Repeat Until <condition>]**. I suggest adding in a ⑥ •**[Start Sound [*Car Vroom*]]** to give an indicator that the race was completed, but that's optional.

Step 7: Adding a Timer
We'll be scoring our race by time, which is a very simple system to implement.

In our *car*, add a new message ① •**[When I Receive [*"Go"*]]** event. This will be used for when the race starts, which will be separate from our •*"GameStart"*, so we can have a countdown (added later). But for now, we'll just add in a ② •**[Broadcast [*"Go"*]]** directly under our •**[Broadcast [*"GameStart"*]]**.

Intermediate Project 4: Big Map Racing

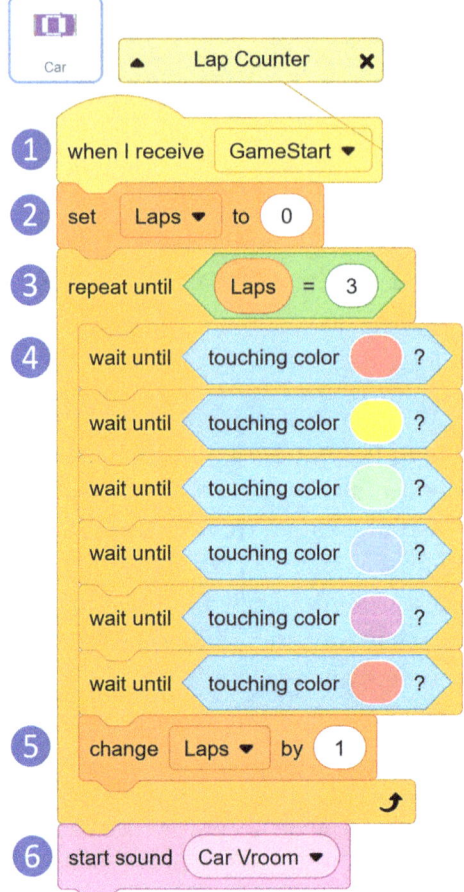

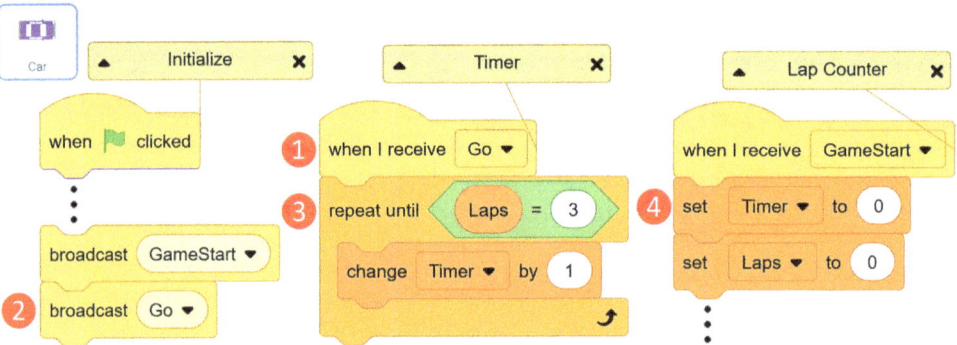

In our *"Go"* event, add in another ③ **[Repeat Until** <("*Laps*") =(3)>**]** loop. This runs only while the race is going on but will stop as soon as the *car* touches the finish line on the final lap. Add a new visible (checked) variable *"Timer"*. This will keep track of the time our player is taking to complete the

Intermediate Project 4: Big Map Racing ◆ 115

race. Add a ④ •[Set ["*Timer*"] to (*0*)] at the start of our •"*GameStart*" event. Then, add a •[Change ["*Timer*"] by (*1*)] inside our •"*Go*" event's •[Repeat] loop. This will run once every frame (30 times a second). It gives us super precise measurement of the time the player takes. As a note, in many timers, you just want to count seconds, so you would add in a •[Wait (*1*) Second] block here as well, but since we want it to be high-precision, we're leaving it out.

Step 8: Driving and Racing States

Our next step is adding in a couple of new states, which will be determined through the two new hidden (unchecked) variables: •"*Driving*" and •"*Racing*". These will help us choose what actions the player can take. •"*Racing*" will determine if a race is actively happening, so we can tell the difference between menu and gameplay, for example. •"*Driving*" will dictate if the player can enter controls for their vehicle, so they can't race forward during the countdown, or when they crash.

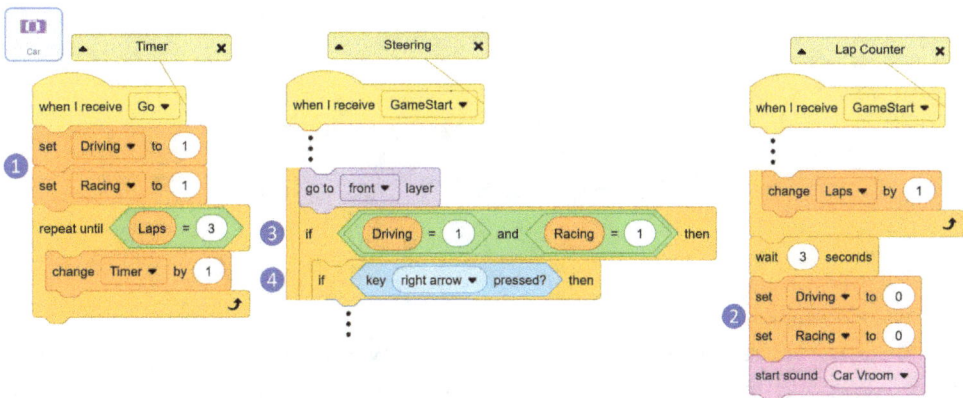

With the variables added, turn your attention to the *car's* code. Just under •[Broadcast ["*Go*"], add a ① •[Set ["*Driving*"] to (*1*)] and a •[Set ["*Racing*"] to (*1*)] for now, so we can still test our game. Under the •"*Lap*" counting •[Repeat], add in ② •[Set ["*Driving*"] to (*0*)] and •[Set ["*Racing*"] to (*0*)], so the player stops controlling their vehicle after they win the race, but also add a •[Wait (*3*) Seconds] above those so they can clear the finish line and still drive for a few seconds, for some added satisfaction.

Next, we'll start using game states. In the *car's* •[Forever] loop, add in a ③ •[If •<•<•("*Racing*")=(*1*)> and •<•("*Driving*")=(*1*)>> Then]. Move the ④ key press •Ifs inside this so that the player can only steer if they are both able to drive, and a race is active.

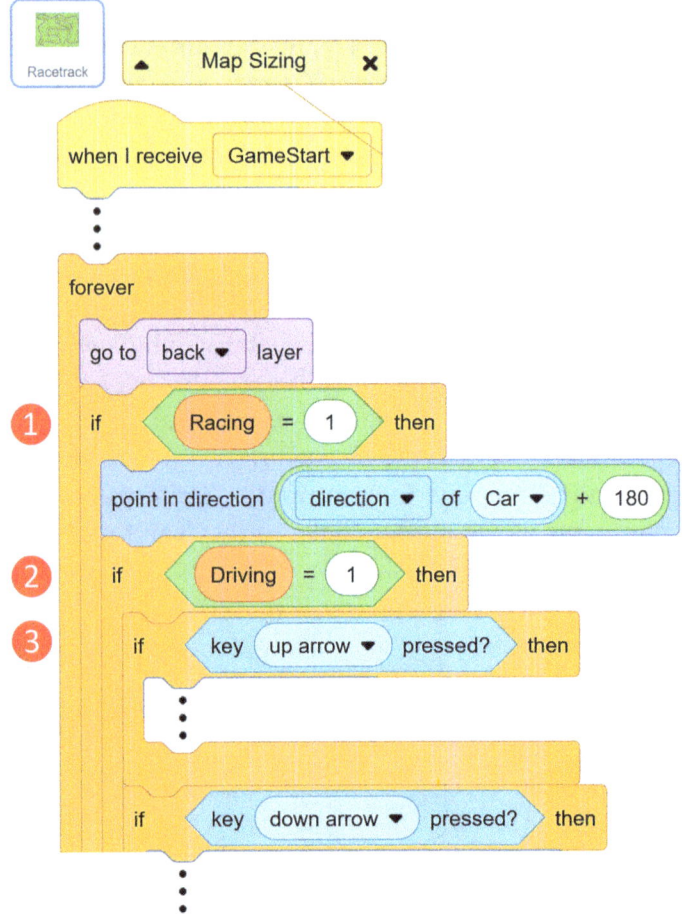

Because we're splitting player control between both the *car* and the *racetrack*, we need to add some code to the *racetrack* as well. Add an ① •[If <•("Racing")=(1)> Then] and place the •[Point In Direction] inside it. Then, add in an ② •[If <•("Driving")=(1)> Then] inside that •[If] and place it under the •[Point in Direction], with the ③ arrow press •[If]s inside it. This way, our map will update whether or not the player is driving, but they'll only be able to move forward or backward when able to drive.

Step 9: Ending the Race

For our next step, we'll add a screen for when the race finishes. Since there isn't a win-or-lose condition in this game, a single screen will take care of displaying your time. Start by adding a blank sprite *RaceDone*, or copy the *racetrack* costume to maintain the same colours used in the race. Make the background a full rectangle of grass green. Across the bottom, place a rectangle of grey for a strip of road, with a yellow/white centre line. For text, we'll add two things. First, a title. Something like "Race Complete" or "Finished!" in large bold letters in the top half. And second, a smaller text display, "Race Completed In:" or "Race Time:", near the bottom. We want our bottom text to line up with our •*"Timer"* readout, so move that in-game variable display to the bottom to coordinate with that text. Lastly, make a copy of our *car*, paste it in the *RaceDone* sprite, and scale and position it on the strip of road.

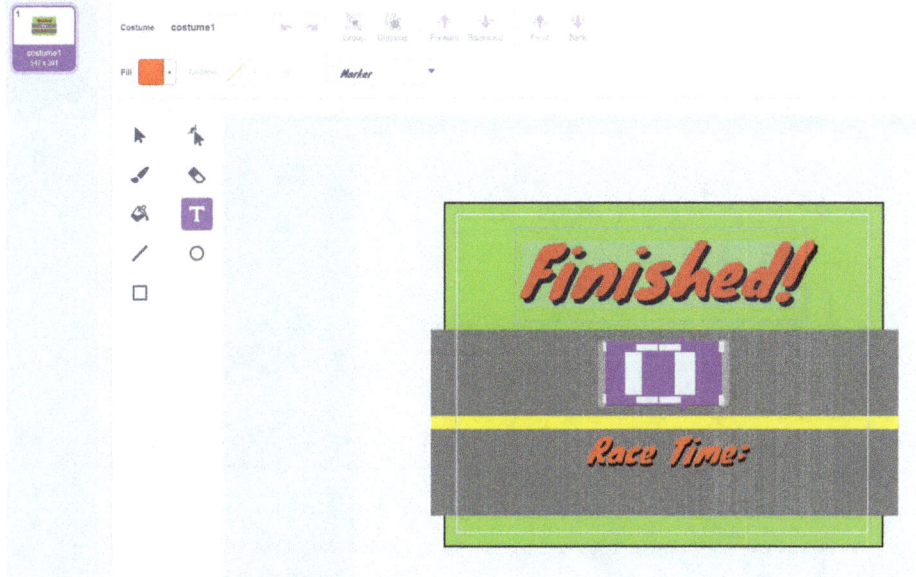

With the graphic done, we turn to code. Start by going to our *car*. Below our lap counter at the very bottom, add a ❶ •**[Hide]** block to disappear the in-game *car*, and add a •**[Broadcast ["*RaceOver*"]]**, used to trigger the *RaceDone* object. Since we've just used a •**[Hide]** code block, just under •*"Go"* we add a ❷ •**[Show]** code block to make sure the *car* will be visible for the race.

In our *RaceDone* sprite, add in a ❶ •**[When ▷ Clicked]**; set it to X: 0, Y: 0, and •**[Hide]** it. In the ❷ •*"RaceOver"* event, add in a •**[Show]** as well as a •**[Hide Variable ["*Laps*"]]**, and a •**[Show Variable ["*Timer*"]]**. We will simplify our display to just the sprite and the one variable that matters at the end: the •*"Timer"*.

Step 10: Barriers and Crashing

Let's add in the hazard of crashing into the game! To start, add a second costume to the *car*. Duplicate the existing costume and use tools like **Reshape**,

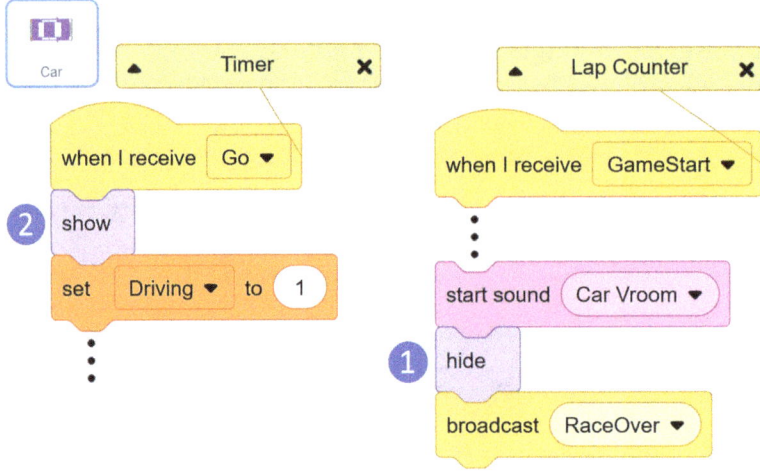

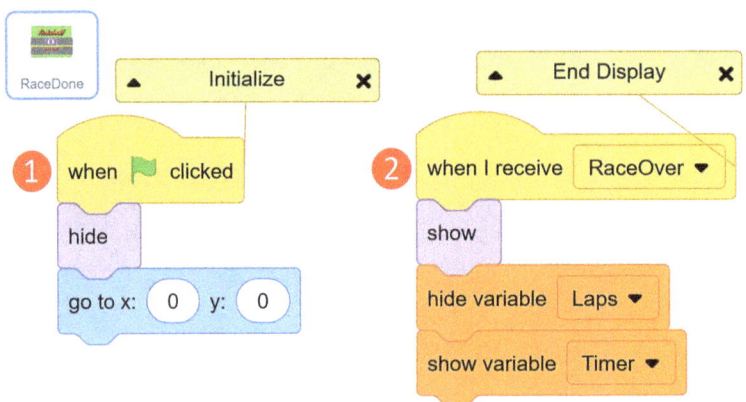

Line, and **Select** to reshape, add cracks, or reposition/twist elements to create an obviously damaged version of the *Car*. You can even add in fallen-off wheel/bumpers or flames/smoke to really make it stand out. Keep in mind how small the car is in the game, so try making them fairly obvious and big enough to be visible.

Next, head to the *racetrack's* costumes. Add in some thin (but not too thin) red barrier lines. These can just be an inner area to prevent cut corners, or you can include some on the outside of the track, or even use them to split adjacent parts of the track, like real hairpin turns. Just make sure you use a single consistent colour of red and make the lines thick enough that they won't be blended into other colours, a minimum of line thickness of 5, to be safe. Make sure you don't use this colour for anything else in the game (like your countdown text in the next step)!

Intermediate Project 4: Big Map Racing ◆ 119

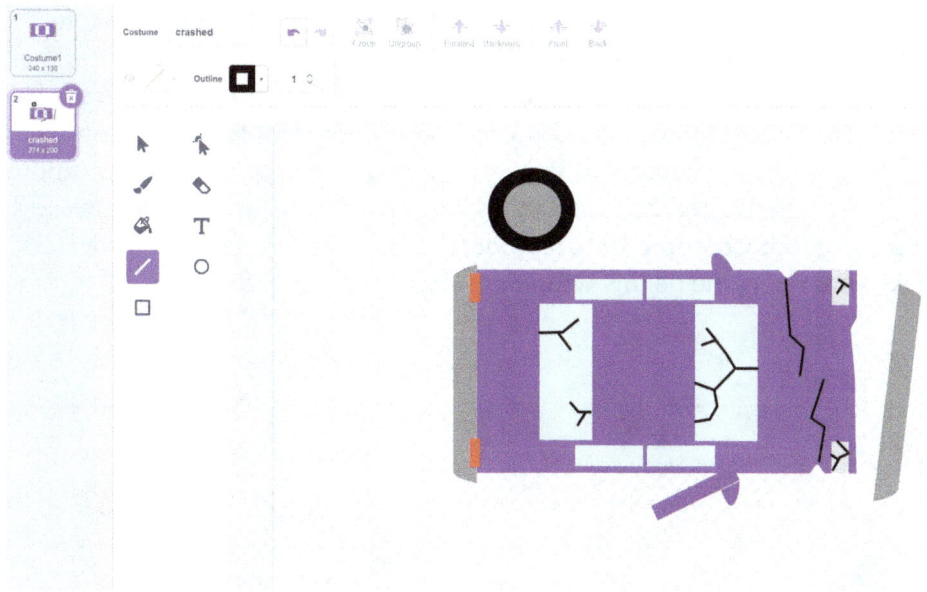

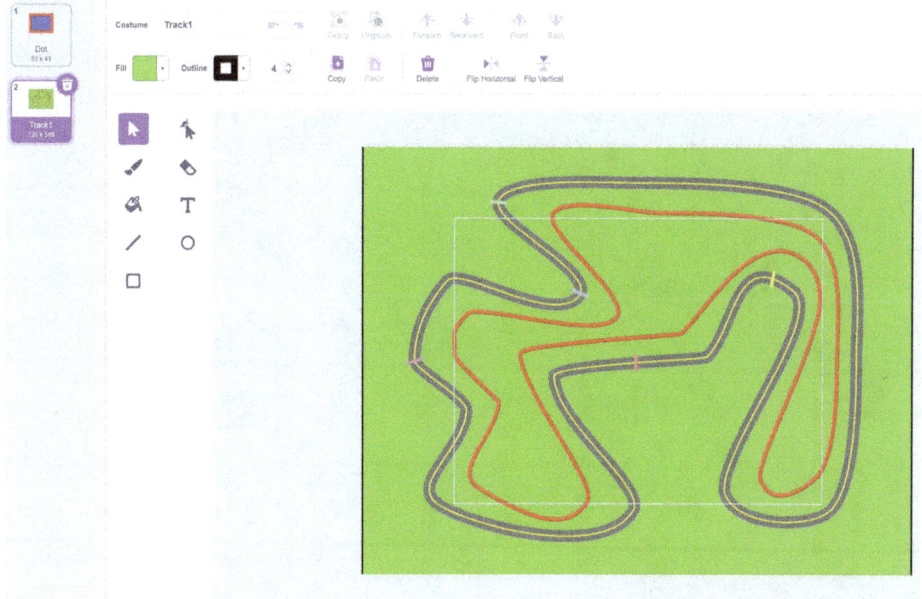

Now, we'll add some crashing code to the *car*. In the •**[Forever]** loop's •**[If** •<•<•(*"Racing"*)=*(1)*> **and** •<•(*"Driving"*)=*(1)*> **Then]** conditional, we'll need an ① •**[If** •<Touching Colour [*Barrier Red*]?> **Then]**, and inside it •**[Broadcast [***"Crash"***]]**. Then, ② create a •*"Crash"* event. We'll ③ •**[Set**

[*"Driving"*] **to (*0*)**] so the player can't drive while in the crashed state. Add a •[**Next Costume**] to switch to the damaged version of the vehicle, then a •[**Wait (2) Seconds**], and another •[**Next Costume**] to switch them back. •[**Wait (1) Seconds**] again, and then •[**Set** [*"Driving"*] **to (*1*)**] to restore control. This way the player will lose 3 seconds (90 on the •*"Timer"*) minimum for each crash during the race. Before we leave the *car*, we should also add in a ④ •[**Switch Costume to** [*Costume1*]] at the start of the //*Steering* stack to make sure the game begins with an undamaged car to drive.

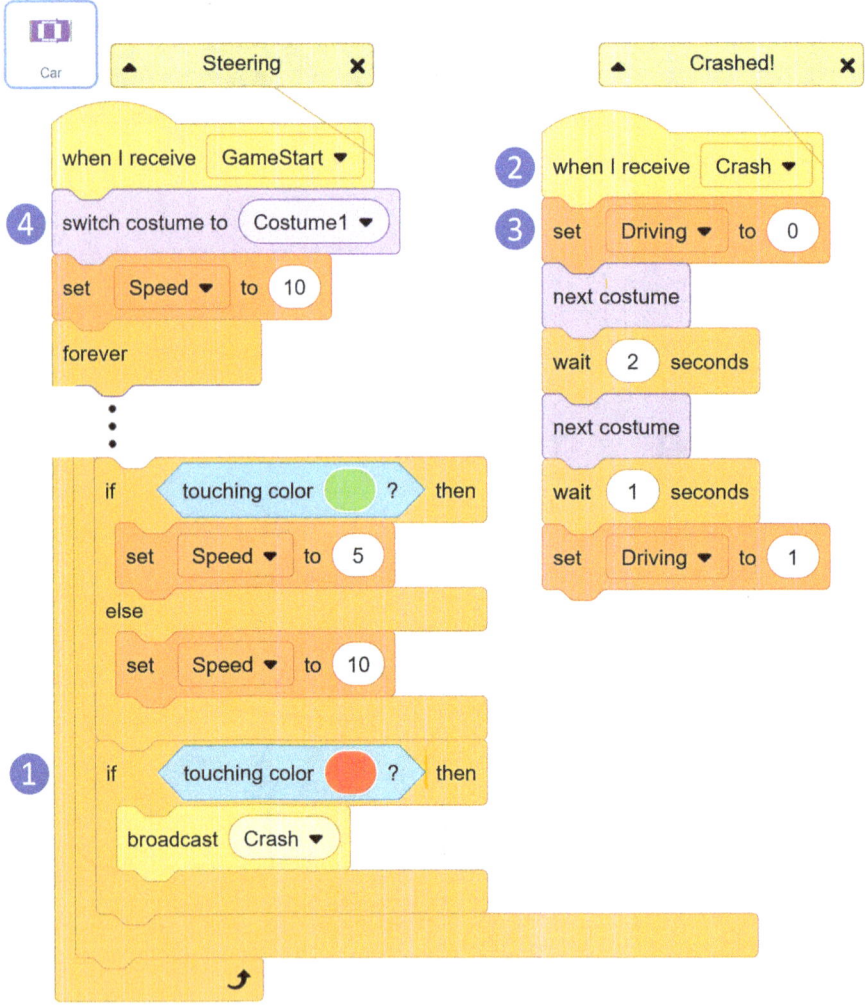

In our *racetrack*, add in a little correction to get the *car* back off any barrier they crashed into. For that we'll need a ① •*"Crash"* event. In it, add a •[**Wait (2) Seconds**], then add a •[**Repeat (20)**] with a •[**Move** •((0)-•(*"Speed"*))

Steps]. This will wait for a couple of seconds for the player to register the crash, then move them backward 20 frames' worth of movement, enough to avoid the barrier this time (hopefully).

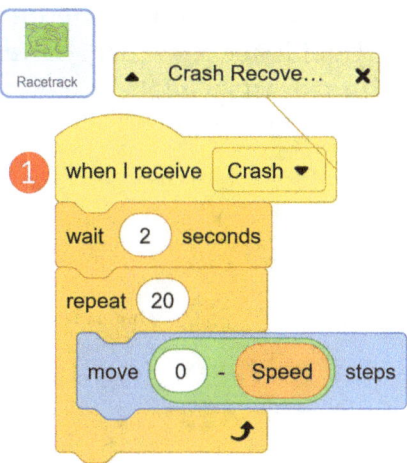

Try it out and you'll see how the car crashes into the barrier, then gets reset back onto the road and repaired before being able to resume driving.

The crash would be an excellent place to add in some more special effects. If you're looking for inspiration, you can check out the Fireworks Display project in Book 1: Beginner or Scrolling Shooter in Book 3: Advanced!

Step 11: Countdown

To place a countdown before the race starts, we'll need a new blank sprite, *countdown*, positioned at X: (0), Y: (0). We can draw a chequered flag or similar design, but we'll want a flat colour background section to allow for some readable text on it, or use drop shadows to ensure the text is legible against that kind of background. Start with one **costume** to say "3", then make copies that say "2", "1", and "Go".

Now we can code our *countdown*. Start with a ① •**[When** ▷ **Clicked]** with a •**[Hide]** block. Then, a ② •*"GameStart"* event. Under this, ③ •**[Go**

To [*Front*] Layer], •**[Switch Costume to [*Costume1*]]**, and then •**[Show]**. This will make sure it starts on the "3" and is in front of everything else. Then, use a ④ •**[Repeat (3)]** with a •**[Wait (1) Second]** and a •**[Next Costume]** inside. This will make sure it counts down switching costumes each second until it's on the "*Go!*" flag. Next, have a ⑤ •**[Wait (1) Second]** (so players can read the •"*Go*" sign), then •**[Hide]** it and ⑥ •**[Set [*"Driving"*] to (1)]**, and •**[Set [*"Racing"*] to (1)]** then •**[Broadcast [*"Go!"*]]**.

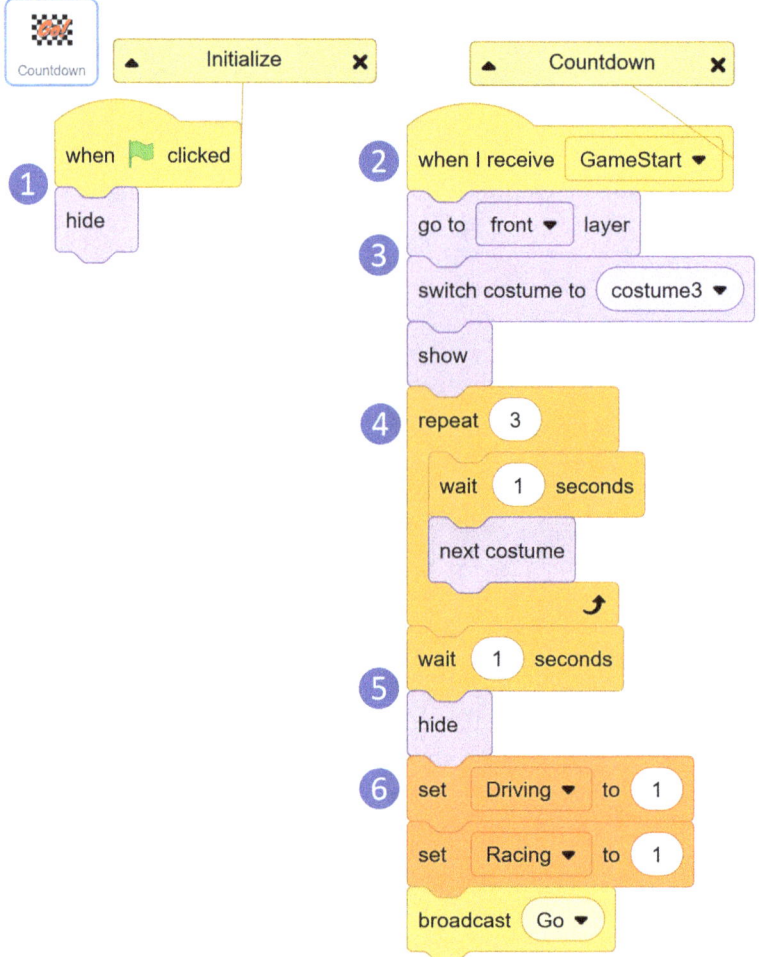

Now that the race starts by the *countdown* object, we should remove a few code blocks from our *car*. In the *//Timer* stack, remove the ① •**[Set [*"Driving"*]]**, •**[Set [*"Racing"*]]**, and from the *//Initialize* stack, remove the ② •**[Broadcast [*"Go"*]]**. Also, let's add a ③ •**[Hide]** in the •**Green Flag** event since we'll be adding a menu soon.

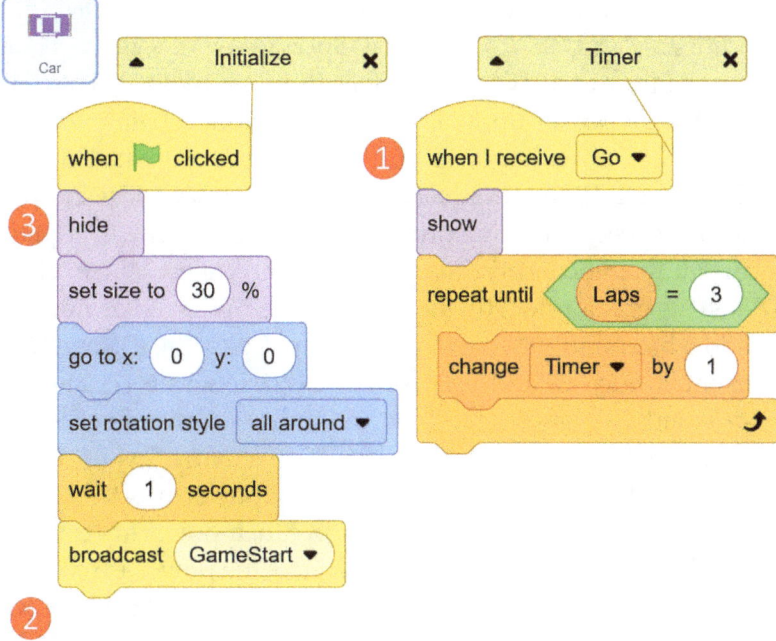

Step 12: Nitro Boosts

For our next addition, we'll deal with a power-up – a nitro boost. For those unfamiliar, "nitro" is a common power-up in racing, modelled on the injection of nitrous oxide into engines to provide extra oxygen to increase fuel combustion and provide more engine power. In games, it tends to give a short-term boost to speed or acceleration. In our game, we'll have it create a 5-second speed boost. For this we'll need two new •**Variables**: •*"Nitro"* and •*"SpeedBoost"*. Hide (uncheck) the •*"SpeedBoost"* variable.

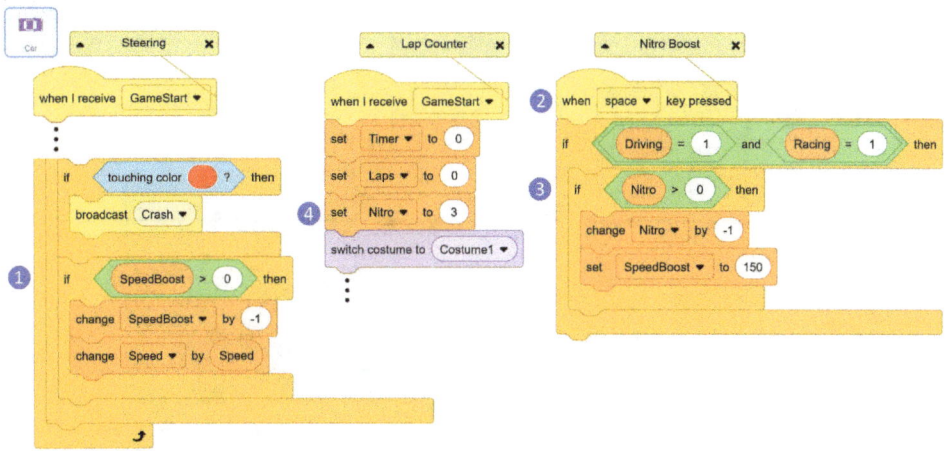

In our *car's //Steering* •[If •<•<•("Racing")=(1)> and •<•("Driving")=(1)>> Then] conditional, add an ❶ •[If •<•("SpeedBoost") > (0) > Then] at the bottom, and inside, add a •[Change ["SpeedBoost"] by (-1)]. This will ensure while our •"Nitro" power-up is active, its duration is being counted down 1 per frame, so our power-up will only last so long. Then, add a •[Change ["Speed"] by ["Speed"]] (alternatively, you could set speed to speed ×2). This doubles the *car's* speed, but being under the off-road test, it will adjust to whatever the current racing conditions are for the *car*. While •"Nitro" applies (i.e., •"SpeedBoost" is greater than 0), they'll get twice the speed as normal!

Next, add a button to activate our power-up. Use a ❷ •[When [*Space Key*] Pressed] event. Under it, add an •[If •<•<•("Driving ")=(1) and •<•("Racing")=(1)>> Then], to ensure a race is on and the *car* is active in order to activate the •"Nitro" boost. Inside, put another ❸ •[If •<•("Nitro")>(0)> Then], ensuring that the player has remaining •"Nitro" boost power-ups to activate. Inside that, add a •[Change ["Nitro"] by (-1)] to reduce their power-up count, and then a •[Set ["SpeedBoost"] to (150)], so once activated, the power-up will run for 150 frames (5 seconds). In the •"GameStart" event, add a ❹ •[Set ["Nitro"] to (3)] at the top to ensure the player gets three power-ups to use during their race.

We didn't get to use too many sound effects in this project, but if you want a few tips and tricks to working with sound, you can check out the Dino Dance Party project in Book 1: Beginner!

Step 13: The Main Menu Screen

Our next step will be to add in our main menu. Start by creating a blank sprite called *MainMenu* and set it to X: (0), Y: (0). A green background, with a grey road rectangle and a yellow strip along the middle. Add a title like "Scratch Racing" and leave the bottom half relatively simple so we have space for some buttons.

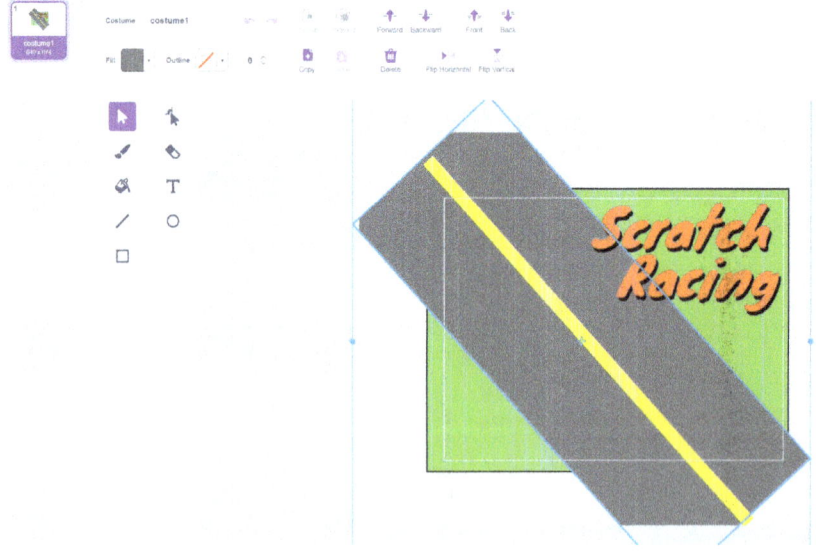

Intermediate Project 4: Big Map Racing ◆ 125

The code for our *MainMenu* will be relatively simple. Start with a ①•[When ▷ Clicked], and under it, put a •Broadcast ["*Menu*"]] event. By doing so, we can reuse the setup code and call it when returning to the *Main-Menu*, not just when we first launch the game.

Under our ②•"*Menu*" event, we want a •[Go To [*Front*] Layer], so the *menu* covers everything else in the game, then •[Show]. Add ③•[Hide Variable [*variable*]] blocks for our •"*Laps*", •"*Nitro*", and •"*Timer*" variables. Also add in a ④•"*GameStart*" event, and in it, place a •[Hide], and add ⑤•[Show Variable [*variable*]] blocks for •"*Laps*", •"*Nitro*", and •"*Timer*". To tie into the *MainMenu*, next go to our *RaceDone* sprite and add in a ①•[Hide Variable ["*Nitro*"]] to our •"*RaceOver*" event, since we added that power-up. Importantly, the player can return to the *MainMenu* after completing a race, so add a ②•[When This Sprite Clicked] event. Under it, place a •[Hide], and a •Broadcast ["*Menu*"]].

In order to play the game when on the *MainMenu*, we'll add a *Play* button. Add a **Button2** from the Sprite Library. In its **Costume** tab, add the word "Race" or "Play". Position it on the *MainMenu* to be below the title on one side of the screen.

To code our *Play* button, start with a ①•[When ▷ Clicked] and attach a •[Hide]. This might seem counter-intuitive, but we want it to activate from the •"*Menu*" event. Add a ②•"*Menu*" event and, under it, a •[Wait (*0.05*)

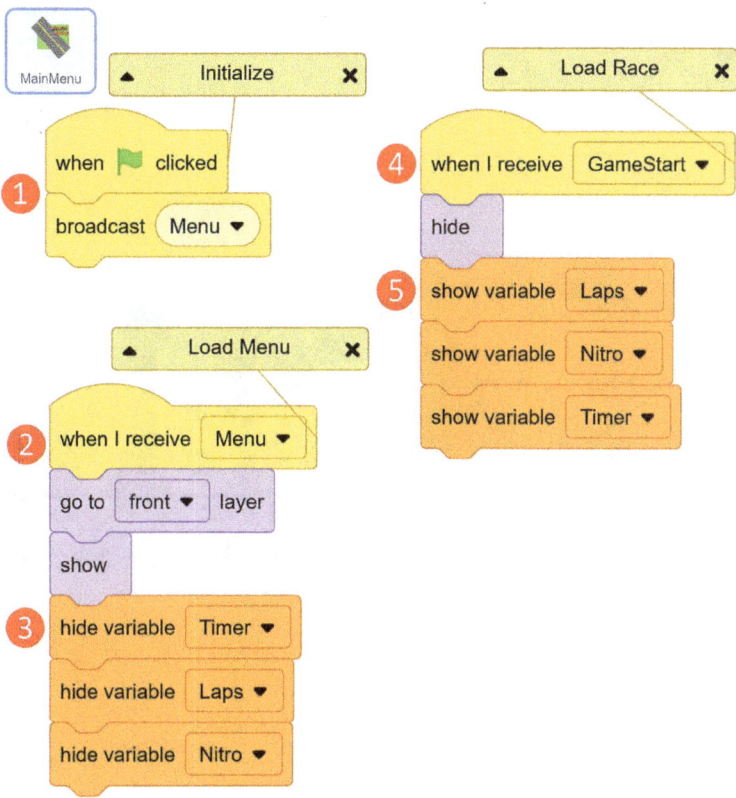

126 ◆ Intermediate Project 4: Big Map Racing

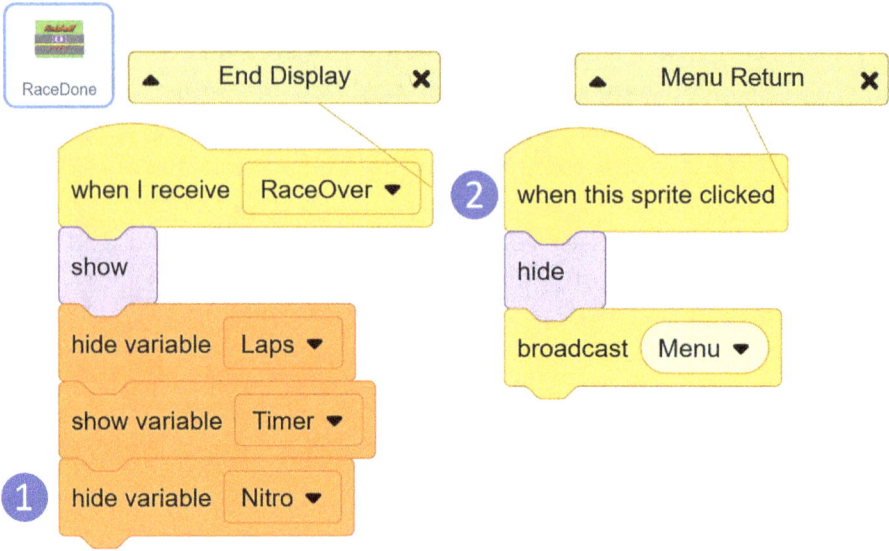

Seconds], •**[Go To [*Front*] Layer]** and •**[Show]**. This will ensure our *Play* button ends up overtop our *MainMenu* sprite, as well as allowing it to be called from more than the launch of the game (like when we finish a race and return to the menu).

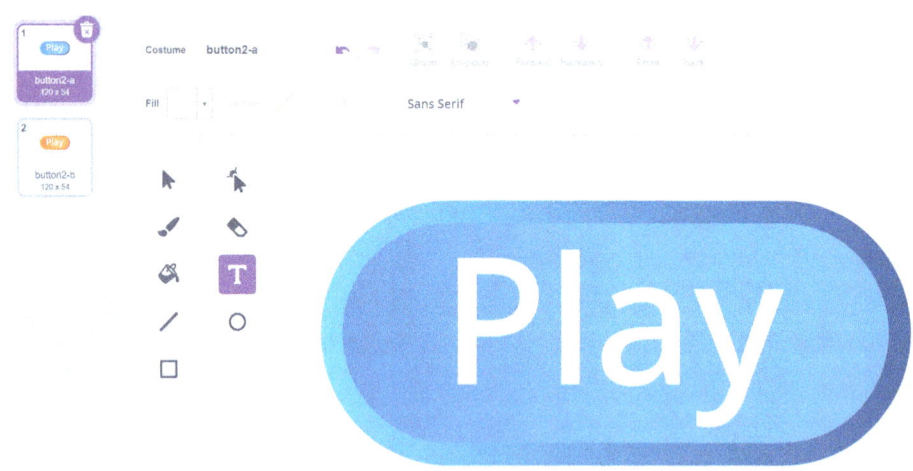

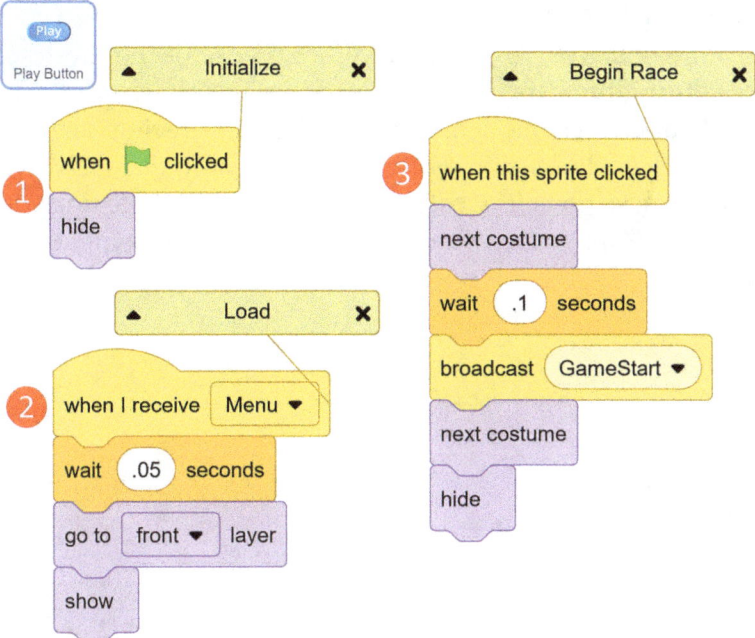

Lastly, add in a ③ •**[When This Sprite Clicked]**. A •**[Next Costume]** to let it flicker to the other costume, •**[Wait (0.1) Seconds]**, then •**[Broadcast ["*Game-Start*"]]**, •**[Next Costume]**, and •**[Hide]**. This will start the game but also switch it back to the default costume before hiding so it is ready to appear again. The very last thing you need to do to make it work is go to the *car* and remove the ① •**[Broadcast ["*GameStart*"]]** code block from the *//Initialize* stack.

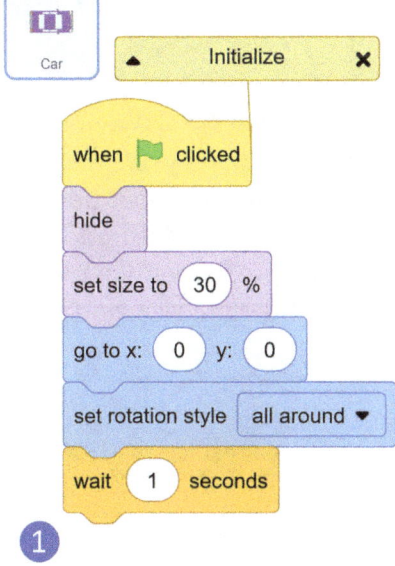

Step 14: Menu Map Display

On our main menu we'll give the player a glimpse of the racetracks in full view. Since during the game they only ever see the blown-up version, it can be nice for them to get some sense of what they'll be racing on. Create a new blank sprite called *MapDisplay*. For the costume of *MapDisplay*, we want to copy the *racetrack*, so go to its **Costumes** and copy the whole track design and paste it into the *MapDisplay*. We don't need the mini rectangle, just the full racetrack.

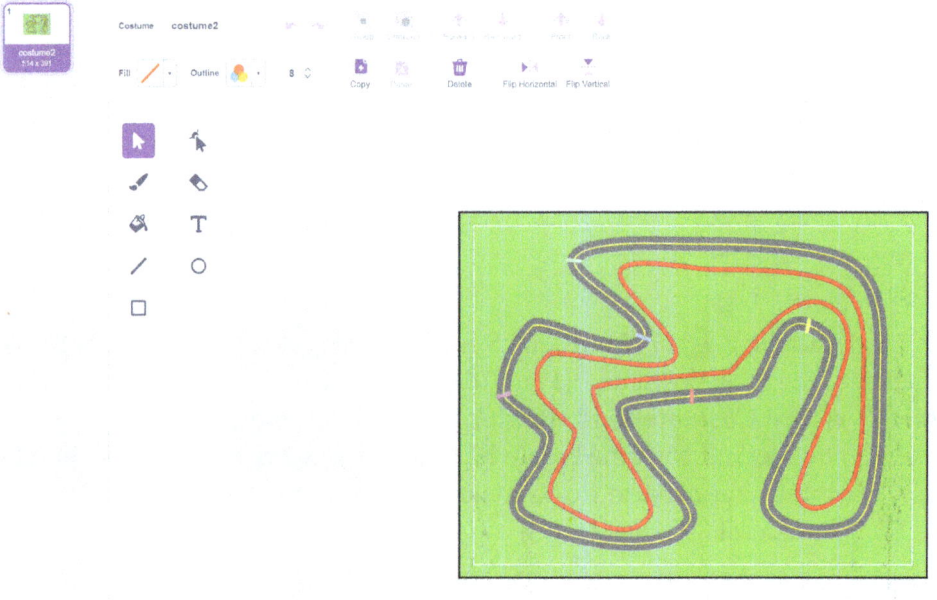

To code our *MapDisplay*, add a ① •[When ▷ Clicked] and a •[Hide]. Then use a ② •"*Menu*" event to display. We'll •[Set Size to (40)%] so it just fills a corner of the *MainMenu* display, then add a •[Wait (0.05) Seconds], •[Go To [Front] Layer], and •[Show]. Also add a ③ •"*GameStart*" event with a •[Hide] to have it disappear when we play the game. Hit the ▷ so you can see the *MapDisplay* at the right size, and position it in the bottom left of the Stage Window.

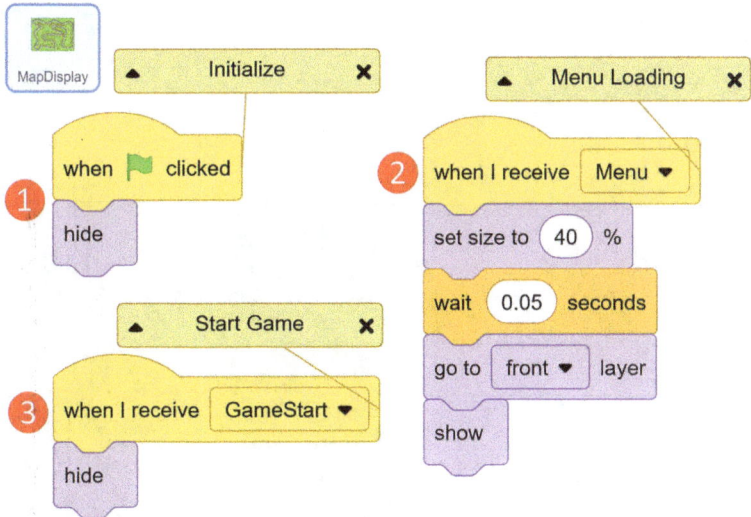

Step 15: Multiple Racetracks

With our race game, we can actually incorporate multiple racetrack designs. To do so, head to the *RaceTrack* and the **Costume** tab. Duplicate the first track, and use the **Reshape tool** to change it around, making sure you reposition the waypoint colour lines too! Make at least one additional track so you can see how to incorporate other tracks, but feel free to add more. Again, if you want to make life easier for yourself, always have the start/finish of the *racetrack* be at X: (0), Y: (0). The *car* and *racetrack* always start there. If you change it, you'll need to add in a system to reposition things to different starts, depending on the map chosen on •*"GameStart"*. Copy the new map into the *MapDisplay* costumes as well so it has the same maps in the same order as the *racetrack*.

We'll need a new hidden (unchecked) •Variable to handle the racetrack selection •*"MapIs"*. Since we've added multiple maps, we now need to ensure our *MapDisplay* shows the correct map. We'll add a ① •*"MapChange"* event to call it from another object, then add a •[Next Costume] and a ② •[Set ["MapIs" to •(•(Costume Number)+1)]. This will ensure that whatever track we've got displaying on the *MapDisplay* will also correspond to the correct costume in the *racetrack* – we have to add 1 because of our mini-rectangle costume in the *racetrack*.

130 ◆ Intermediate Project 4: Big Map Racing

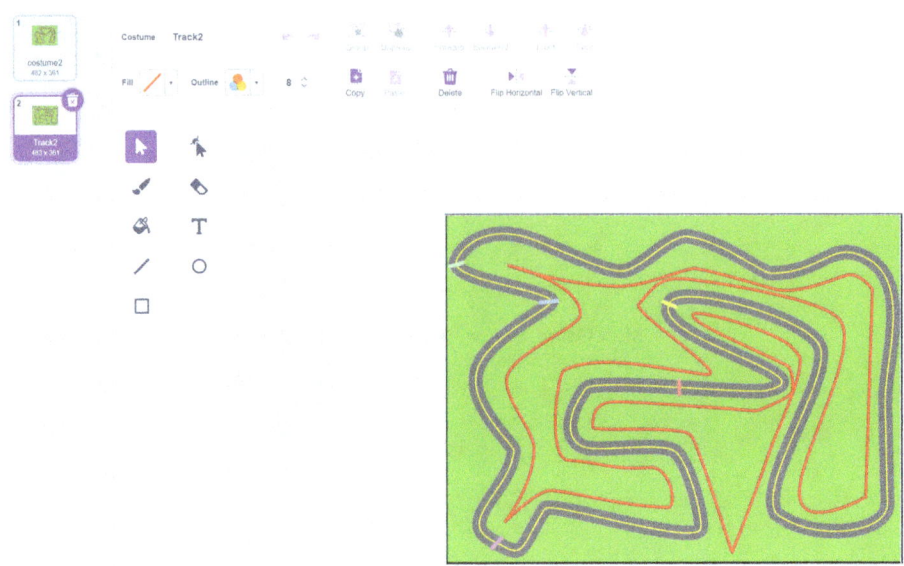

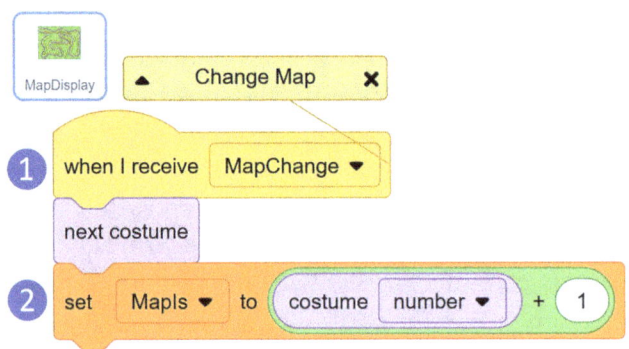

Next, add in a *Map button* to switch the map. We could allow the player to click on the *MapDisplay* to change tracks, but it might not be obvious. So instead, duplicate our *Play* button, and in the **Costume** tab change the text from "Race"/"Play" to "Map" or "Track". Then, in the code, in the •**[When This Sprite Clicked]** code we just need to change its event from •*"GameStart"* to ① •*"MapChange"* and remove the •**[Hide]**. Now, clicking this button will cycle through our possible tracks and have the *MapDisplay* update to the new track. Add a ② •*"GameStart"* event to •**[Hide]** the button.

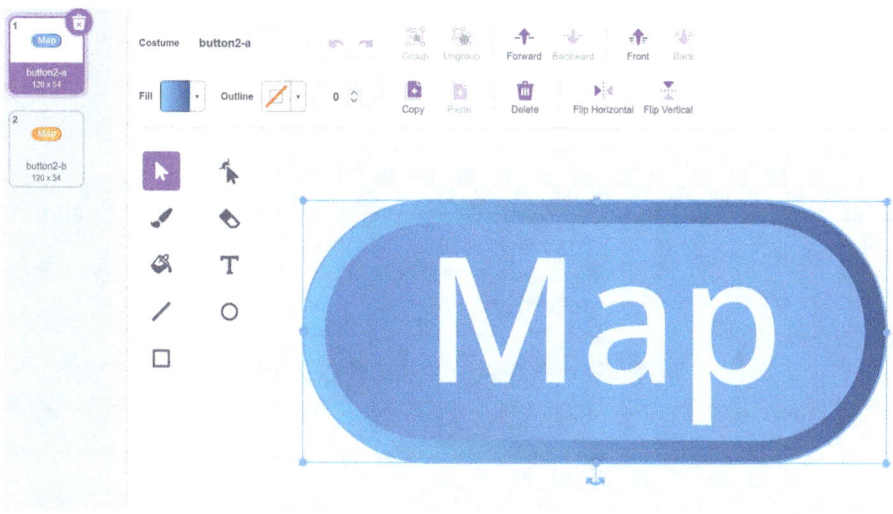

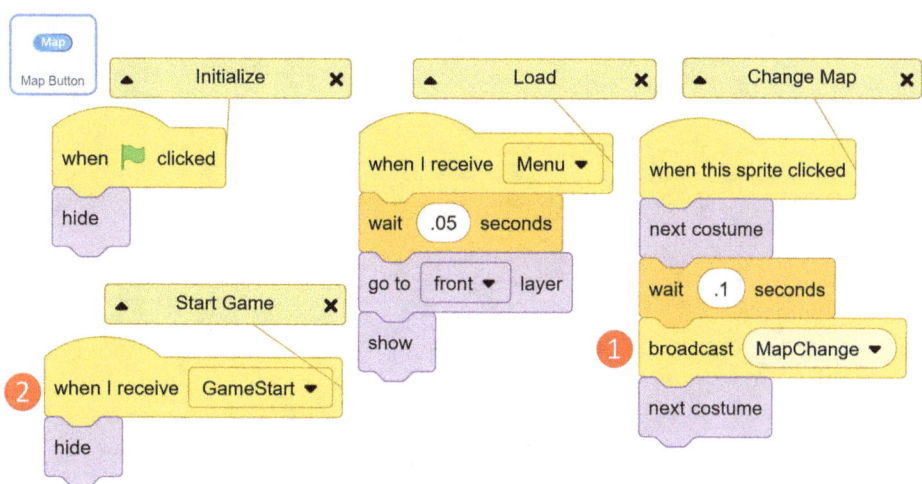

Updating to our chosen track requires changing the code in our *racetrack*. All we need to do is find the •[**Switch Costume to [*Track1*]**] after •[**Set Size to (*1000*)%**] and add in a variable •("*MapIs*") block so it reads ① [**Switch Costume to** •("*MapIs*")]. This will ensure the map switches to the map we selected on the *MainMenu*.

Lastly, add in some track-specific custom starts if needed. To do so, in our *car*, go to the •"*Go*" event. Add in an ① •[**If** •<•("*MapIs*")=(2)> **Then**] and a •[**Point in Direction (45)**]. This is an example, showing how to custom-orient the *car* to the appropriate direction based on the map. Depending on your track designs, you may want the *car* to start pointing in any given direction, and this little technique will take care of that. Alternatively, you could

132 ◆ Intermediate Project 4: Big Map Racing

use a similar system to dynamically position the *racetrack* if you don't have X: (0), Y: (0) as the starting point for each track.

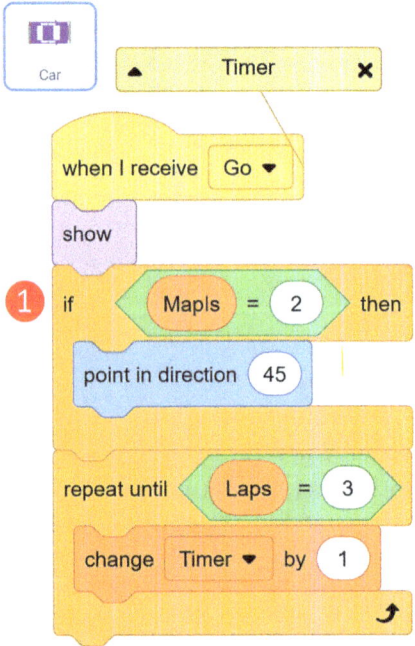

For working copies of this and every project in the book series, visit www.massivelearning.net *for direct links to Scratch projects, and to see our other projects and resources for coding education!*

9

Intermediate Check-In

Now that we've completed our four intermediate projects, let's stop and reflect on our progress.

Key Skills

These four intermediate projects added a lot of scale with touches of polish and consideration, rather than the bare-bones, just-get-it-working approach seen on beginner projects. You've become familiar enough with Scratch to not need the direct hand-holding to find blocks (except maybe a few hard-to-understand ones), and you should now be able to operate pretty quickly and seamlessly building out code. At this point, you are probably pretty good at predicting a lot of basic patterns, like how and when to use comparison Ifs or create messages to trigger things. We've used these four projects to show you four very different kinds of projects. We started with a drawing program, introducing extensions and highlighting geometry. Next, we switched things up with an interactive story, focusing on storytelling, perfect for English classes, while also teaching the basics of messages and text handling. Then, our Snowball Fight focused on more physical games, showing how to use clones as well as model movement and collisions, fit for discussions on the science of physics while exploring some math with angles and iterative functions. In our Big Map Racing, we created a large-scale map and cars that allowed us to flex our art and creative muscles, while the code let us learn more about working with variables, inputs, and conditions, alongside some great menu features. We've deepened our knowledge through learning more advanced techniques, which expanded the scale and conditional use of the simpler systems we were already familiar with.

Let's think about some of the core skills learned and practised with these projects.

Events and Triggers

Code only runs when it's attached to an event, triggering it. It's one of most important elements in Scratch. We expanded on the basic events by learning to work with not only a few more unusual ones but also learned some more twists and turns that can add to the flow and sequence of how code runs.

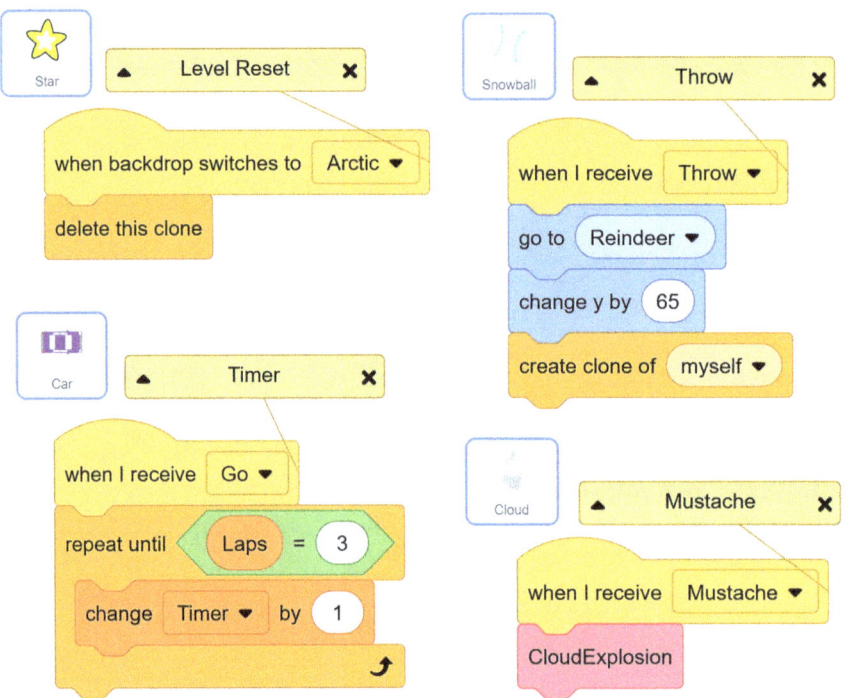

Figure 9.1 Some examples of how Events and Triggers were used in projects. The Star triggers on a backdrop switch. The Car receives a message. The Snowball receives a message and creates a clone event. The Cloud triggers on a message and calls a My Block event.

Messaging

We made very extensive use of the messaging system in Scratch, creating our own custom events and calling them to organize our projects' flow.

Through the use of messages, we learned to create distinct scenes in our stories, menus separate from gameplay, different levels to play through, or even interruptions to play, such as crashes in a race. Messages have a variety of uses and can be as simple or complex as the project requires it, enabling the creation of custom events to be called in specific moments, as needed.

Custom Blocks
Complex functions can be built and run with the use of custom •My Blocks whenever we needed, without repeating code. We learned how we can use this for custom FX in our interactive story and handle common processes like ending a game turn in our Snowball Fight.

Until Loops
This new form of loop enabled creation of interesting new gameplay controls. •**[Repeat Until <*condition*>]** ensured snowballs could fly through the air and end when they struck the ground, or timers could run until enough laps were run. We also learned to hold processes, ensuring proper sequencing, with the •**[Wait Until <*condition*>]** block creating a specific set of waypoint collisions to occur for counting laps in our Big Map Racing.

Dealing with Scratch Limits
With these projects we started to rub up against limitations in Scratch. As a platform made to be as beginner-friendly as possible, they added a number of behaviours that the code itself might not reveal. In most coding languages, objects can simply move through an infinite plane, and variables or properties can have nearly limitless values. But Scratch will intervene to prevent some possible problems for its users. These can get in the way of things we want to do.

Size Limits
While trying to make our Big Map Racing, it became evident that we only have so much room to work with. Objects can't be expanded too large or shrunk too small. Scratch wants to make sure users (mainly kids who are new to coding) don't get stuck, unable to see an object they created, or make something so big it obscures everything else. It's a reasonable thing to do when mainly dealing with such young users, but it can be a pain when it stops you from being creative. We learned about using alternate small (or large) costumes to bypass the protection, having a background as large as we wanted for our level.

136 ◆ Intermediate Check-In

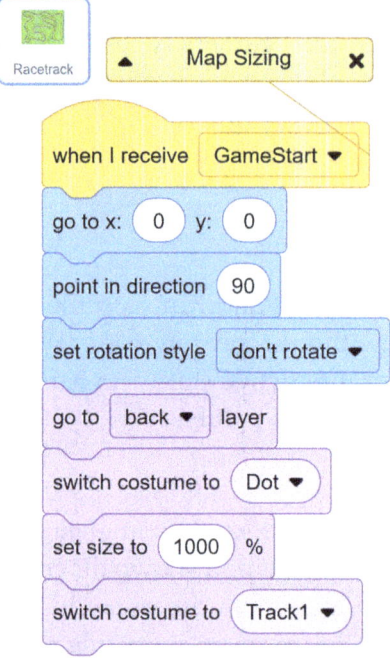

Figure 9.2 An example of the size limit work around from the Map sprite in the 4th project – Big Map Racing. We switches the costume to a very small one, increased the size and then switched its costume back to the large track bypassing the size limit!

Position Limits

Like size, Scratch protects (limits) the position of sprites. It prevents things from moving off-screen by making their maximum position based on still having some portion of their sprite visible on the Stage Window. This ensures that Scratch users can still click on and drag any sprite they've added to the game and not lose them off the edge. With loops, it's very easy to have things move off-screen without this protection. A reasonable thing to have in place that requires use of some special techniques to get around, like with our racetrack.

Variables

We learned a lot of new tricks to use with variables. Variables power almost all the amazing things we can do with computers, since they're math machines made to crunch data, and variables are the biggest part of data. We saw them put to use in lots of new, very reusable techniques, giving you a ton of tools to work with when creating your own projects. We also learned a few more things about working with them in the Scratch editor specifically, like using sliders.

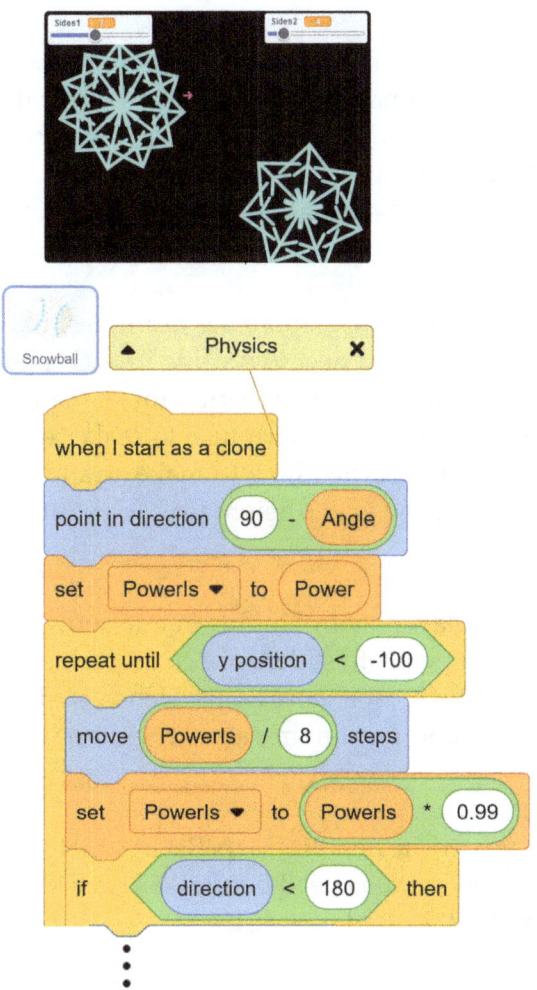

Figure 9.3 Some important variable techniques. At the top the variable slider user controls from Project 1 – Pen Tool Fun. Bottom working with variables for physics effects in Project 3 – Snowball Fight.

Math Altering

While we dealt with the •**[Change Variable by (#)]** code block to alter our variables in the beginner projects, now we started working with math functions to alter our variables in more complex ways than simple adding or subtracting. In our Snowball Fight, we used iterative functions to scale our movement rate with a multiplier, ever slowing in the face of air resistance. We saw how we can set a variable using math modified to modify its value. We applied variables through fractions as *wind* on our snowball's flight with a transform, and our *angle* variable with an offset from the Scratch default values of direction.

Timers
For the Big Map Racing, a timer system was used to measure our race performance. While we could have created a second-accurate count by using a **[Wait (1) Second]** code block, instead we used a frame-rate timer counting in 1/30th second intervals for higher accuracy. We also made countdown timers for power-ups, so our nitro boost lasted only 5 seconds in the game.

Randomizing
We got to apply a number of randomizing examples, like setting our wind speed to a random positive or negative value in each bout of our Snowball Fight. We created a random starting position for our sprites in the Snowball Fight, using a more complex math formulae to make them position in intervals of 25 pixels. We even learned how to make a randomizing function for costumes to switch the snow wall's height for each bout.

Limiting, or Clamping
By using variables as player control inputs in our Snowball Fight project, we limited the possible ranges with sliders of 1 to 100 values, but with keyboard inputs, players could still escape those limits. Hence, we learned to put in tests and enforce minimum and maximum values to clamp values to a set range.

Economy or Usage Limits
Power-ups were created in our Big Map Racing that players could use to give themselves a time-limited speed boost. We also tested to make sure they could only engage it if they had uses left.

Hiding or Showing
Variables can be hidden or displayed to clear the screen for Menu or Game Over screens but still provide the useful information during gameplay.

Object Referencing
We learned to use a very powerful but obscure code block: **([Background #] of [Stage])**. Through this code block one sprite used another's properties to set its own behaviours or properties. In Big Map Racing, it ensured both the car and racetrack were aligned to the same angle of movement.

Inputs
Handling inputs is a key skill to interactive media, and improving your grasp at handling input to create or handle complex conditions can help you make deeper experiences and more complex simulations. We learned a few new ways to get, interpret, or handle inputs, either from the user or testing states in the project.

Intermediate Check-In ◆ 139

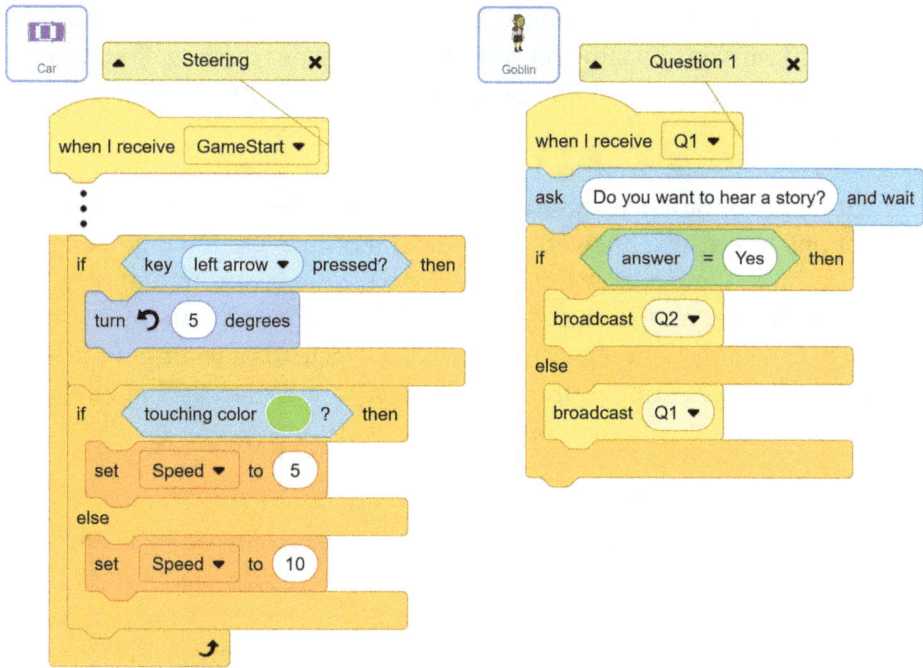

Figure 9.4 Three useful input controls from the projects. On the left the Car sprite from Project 4 – Big Map Racing uses both keyboard control inputs as well as colour sensing tests for its movement. On the right the Goblin in Project 2 – Interactive Fiction uses the Ask block to elicit text responses from users

Key Press Tests

While we used **[When [@] Key Pressed]** events in our beginner projects, here we learned to handle key presses without their own events. By using **[If]** statements to test the condition of a key, whether it is pressed or not, not only more responsive systems can be created, but we could also chain them together, so two keys needed to be pressed at a time, or determine if a key is pressed within another event or loop. This technique allows us much more reactive controls and can be handy for creating combination controls.

Text Input and Saving

In our Interactive Story project, we learned to get text responses from our player with the **[Ask]** code block, as well how to test these answers, and the difficulty of predicting human behaviour. We employed some handy techniques for handling human input and how to save answers for use throughout a project.

Colour Collisions

For our Big Map Racing project, we created a number of uses of colour collisions using the **<Touching (Colour)?>** block. This allowed us to interact with our backdrop to create off-roading effects, barrier crashes, and waypoints to ensure our players were completing full laps. We also learned about the dangers of blending/anti-aliasing on colour accuracy, and the need for minimum width and size to ensure colour collisions will work properly.

Logic Connectors

In our Interactive Story, we used ●<<> Or <>> code blocks to connect multiple options to test text answers against. In the Big Map Racing, we learned about two more logic connectors: "And", to ensure two different conditions were met to trigger an If condition, and "Not", to check for an inverse condition, such as testing if there wasn't a collision, instead of testing if there was. The logic connectors or operators can be very powerful tools that allow us to get far more specific with our conditions while running code.

Game States/State Machines

We made a lot of use of game states in our intermediate projects. While we touched this earlier, we took the concept and really broadened it out to incorporate a lot more common uses. With game states, we learned to make a player turn system, control lockout, intro scene, main menus, and Game Over screens. Because not all projects are games, the term "state machine" is used as a computer science term to encompass this concept more broadly.

Game Turns

In our Snowball Fight, game state systems were used to alternate between player and opponent turns. Using both ●Variables and trigger ●Events, the player and opponent traded throws at each other. This same system could be used for multiple human players or to simply divide play into multiple turns in a cooperative game, or one without opposition.

Control Locking

In our Big Map Racing project, we used the *"racing"* ●Variable to determine if a race was currently running so we could separate out menus from gameplay. A *"driving"* ●Variable determined whether the player should be able to control their car. This was handy not only to limit controls during gameplay but also to lock them out when a player crashed the vehicle. Controls are contextual, so having game states to control when and how they apply is a very useful concept for almost any project.

Levels and Tracks

In our Big Map Racing, we learned how to add in multiple racetracks. By using a ●Variable, we could easily change what level or race the player was on and had some simple game state controls to adapt to that selection. We even

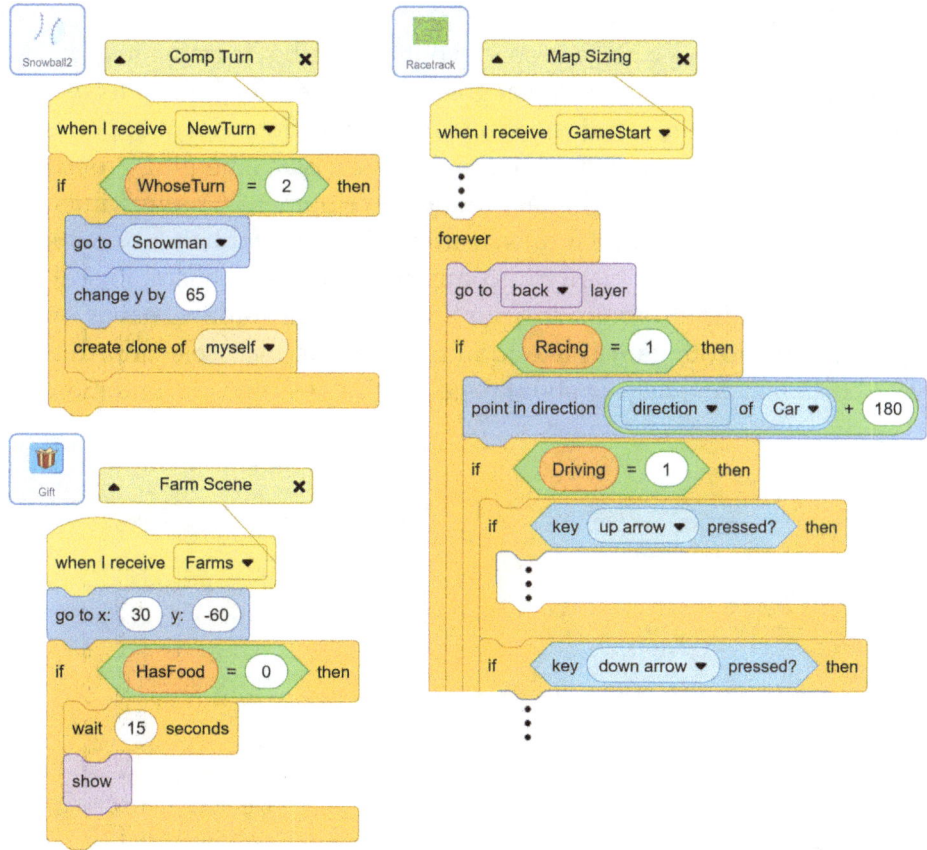

Figure 9.5 Three examples of state machines, Snowball2 showing part of the turn order system from Project 3 – Snowball Fight. The Gift from Project 2 – Interactive Fiction using a variable test that can prevent duplicate spawning so players pick up items only once. The Racetrack from Project 4 – Big Map Racing using the Racing and Driving variables to determine what updates and controls should be available based on the game state.

added in menu interface elements to allow players to choose levels and get visual feedback about their selection.

Event and Stage Switching

In our Interactive Story project, we used custom message Events to control the flow of the story and what scene we were experiencing. There are lots of ways to use messages and stage backdrop changes to control the game state and trigger code. Some important transition and exit handling in the project were applied to also make sure everything in a scene disappeared when the scene ended before the next could load. Remember that we can't just deal with starting our scenes; we have to make sure we exit them too!

Building Interfaces and Menus

It's always nice to have a title screen for your project, but learning to go beyond just the absolute basics can be a real great skill to develop. While focusing on gameplay is a lot of fun, learning to build interfaces offers a lot of opportunities that might not be obvious. UX, or user experience, is an emerging professional field that lies at the heart of digital product development. Through interfaces, we as designers speak to and influence user behaviour, for better or worse. It's important to think about clarity of communication, symbolism, logical flow, and empathy, while also polishing our art skills. By including menus, we can add more depth to our games, like adding in level, or track, selection, difficulty levels, customization, help or about pages, high score tracking, and more.

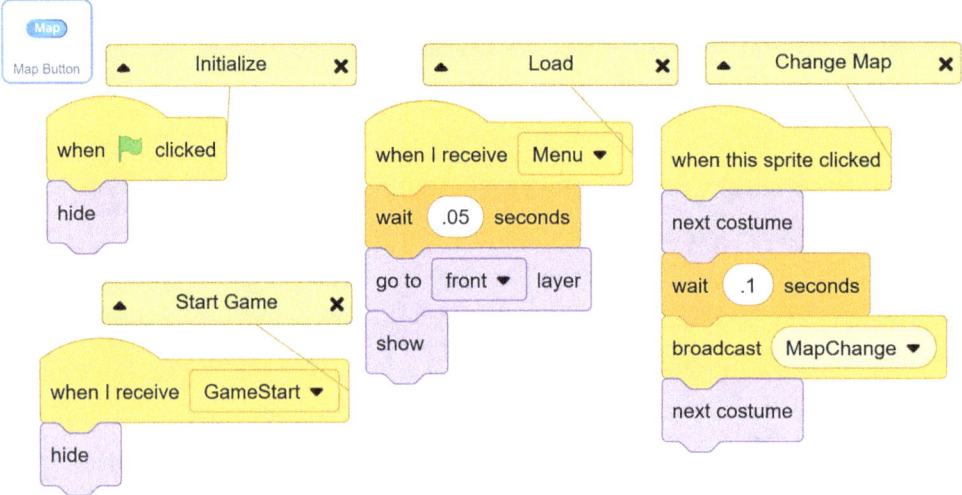

Figure 9.6 The Map Button from Project 4 – Big Map Racing. This technique gives a hidden element that only shows when called, ensures it's on top of all other game elements and gives a nice button press effect when clicked.

Messaging Calls

Messaging is an enormously valuable tool for creating interactive objects and for timing complex events in Scratch. This is especially true for interface elements, who almost always need to trigger other things to react to user selections and input. We used buttons and other inputs to trigger messages that made other objects adapt in response.

Layering Elements

To build an interface, one will need to layer objects so that buttons are in front of the backdrop at the very least. While the •Looks code blocks •[Go To [Front] Layer] and •[Go [Forward] (1) Layer] are useful tools, we learned how to apply them through messages and wait blocks to ensure more complex layering of multiple objects can be achieved correctly.

Variable Setting Buttons

To choose the racetrack in our Big Map Racing project, we created two interacting objects that set •Variables. We used a button to cycle our map display so the player could see an accurate map of the racetrack they would race on, but we also used that to set an internal variable to correctly load the level when they chose to play the game. Having menu buttons to change settings is a very handy way to create adaptive experiences, which we'll explore more in our advanced projects in Book 3: Advanced.

Clones

One of the most powerful features of Scratch, and one of its more confusing ones, are clones. We worked a lot with the clone system in our Snowball Fight project. This gave us a taste of what the system can do. Being able to dynamically create sprites during gameplay and having multiple instances of sprites without dealing with them as completely separate sprites in the game project are very convenient features of cloning. Understanding cloning can take some practice, because there are some things that can easily hang people up, but it's a very valuable skill in Scratch, and a key concept when moving to other computer languages.

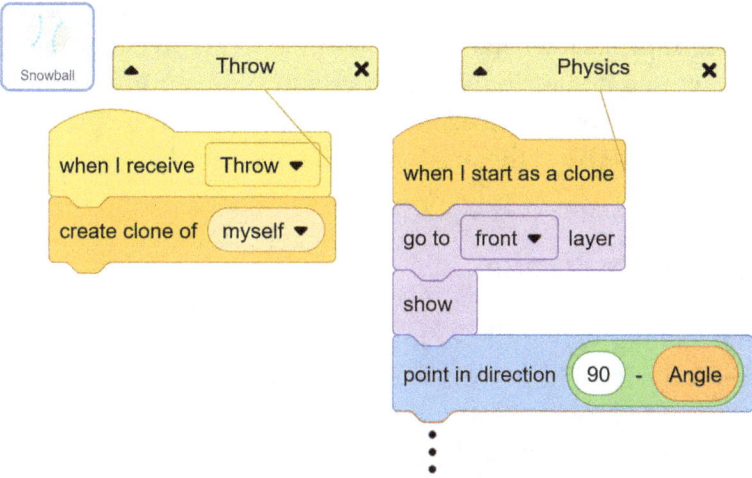

Figure 9.7 An example of how Project 3 – Snowball Fight uses clones to create projectiles. The Invisible Snowball sprite is used to create clones of itself that move and react in game.

Temporary Objects

Our first use of clones was to make a snowball that we could track and then destroy at the end of a turn in our Snowball Fight. Clones can be a great way to create temporary versions of an object. This can be great for things like projectiles and other destructible or consumable objects but can also be used for special effects like mouse trails or particle special effects.

Multiple Objects

In our second use of clones, we created multiple copies of the bonus star in our Snowball Fight project. By cloning our star, we could offer the player multiple possible bonuses to earn in the level. While the objects in this case were temporary, it was more about having multiple copies of the same object that we could place throughout the level, all with the same interactivity and coding, without needing to make them set objects in the game that we always had to handle. This keeps our projects simpler and cleaner, with minimal sprites to deal with, and no duplicate code between duplicate objects. As a bonus, if a bug is found, it doesn't have to be corrected in multiple places.

Original vs. Clone

While we didn't do a whole lot with clones, we did learn a very important lesson to working with them – the difference between the original and its clones. The original instance of an object can receive the •[When ▷ Clicked] because it exists even at the start of your project, whereas clones are always created during the game and therefore never receive the •[When ▷ Clicked] event. Clones get the •[When I Start As A Clone] event, but the original never receives it. This means that the original and the clones will act differently. We must always plan for this unusual arrangement. In our Snowball Fight project, we handle the original as an invisible controller that handles spawning clones that are the actual visible and interactive component in our project. We also learned that clones inherit the properties of the original, so if we •[Hide] the original, we'll need to •[Show] the clones.

More Intermediate Practice

We've come a long way in these four projects. Their scale and complexity both increased a lot. I hope you're feeling good about the projects, even if some of the code blocks or combinations might have been a bit of a surprise at first. This level of programming covers the majority of the projects on Scratch, so hopefully looking at other people's projects with the See Inside button is starting to make sense. If there's a lot of custom •My Blocks and •Variable code blocks, it's a sign it's an advanced project, and no worries if you can't follow those just yet. It's a good idea, even if you are feeling confident at this point, to practice a lot before moving on to the advanced projects. Get comfortable, use the methods we've seen, and try building your own versions of projects from zero without the See Inside button. Here are three ideas for more projects you could try to make.

Infinite Swimmer

Infinite Swimmer, or Infinite Scroller, is a game where the player controls the lateral motion of a character, but they automatically progress through the ocean, or space, or whatever, and must avoid obstacles. In this kind of game, you'll want a highly mobile character, and you'll need a system to create obstacles, and possibly pickups, that will move to create the illusion of constant motion. You'll need to develop a system for creating new obstacles and pickups, moving them, testing for collisions, and destroying them after they pass off-screen, behind the player. These can be themed in any number of ways, but with Scratch's default graphics, underwater and in space are very popular. This kind of project will help you practice working with clones, responsive movement, and collisions, but you can always add in menus and other features. We'll show you an advanced version called Scrolling Shooter game in Book 3: Advanced, but you should be able to make a simpler one with your current skills.

Interactive Diorama

One of Scratch's great uses in education is for students to create interactive subject reports. They can not only present information but also bring it to life with animations. Because the space is so limited in Scratch, you need to come up with ways to hide or display information as needed. We've dealt with visibility, but can you create a page system or create scrollable text? Can you illustrate subjects with animations or even interactive animations? There are a lot of great diorama and report projects on Scratch. You can find ones about animals, science principles, cultural practices, historical topics, and more if you need some inspiration. Think about the methods you use and you'll be ready and able to help students discover and understand the systems they might need for making their own projects.

Pong

A classic video game inspired by the table tennis–based game. Players (or one player and one computer opponent) control paddles on opposite sides of the screen, and while a ball bounces around, they must stop it from hitting the wall behind their paddle by intercepting it to return it to their opponent. A missed ball provides the other player with 1 point. For creating the Pong game, you'll need movable paddles and a bouncing ball. There are numerous ways to create a bouncing ball, so see what you come up with. Perfectly accurate physics is nice to be able to create, but it may not give the best gameplay. You can add a lot of tweaks to the game to keep it interesting, like changing the speed, size, or behaviour of the ball or paddles. Is your game for one or two players? Can you create a project that allows you to choose between the two? This project can be as simple or as challenging as you want to make it.

Teaching Intermediate Scratch

Our intermediate projects have expanded our skills and understanding of coding a lot. Our beginner projects were easily suited to grades 2 to 5 students, but these projects require a level of sophistication and complexity better suited to middle grades 6 to 9, as they expand our skills and understanding of coding tenfold. Whatever grade you teach, it's still important to understand the larger picture of coding and have a deep understanding of the tools and methods to best prepare students. We want to be guides that can describe what's ahead, beyond the path we're walking with them. By learning more about the underlying principles of computer science, we don't just learn one tool, one project, or one assignment, but we go beyond that: understanding concepts that don't just underpin coding, but the universe itself in logic and information theory. Tools and practices will come and go, but by learning the fundamentals, one can easily adapt to whatever the needs of the day are. By learning the principles, we can have the knowledge provide success maps for students beyond our classrooms, having the confidence to handle disruptions, changes, and challenges that will inevitably come. Let's look at some of the things we can bring to our teaching practice when we become confident intermediate coders.

Working with Scratch

Understanding the platform is one of the biggest hurdles. By completing the beginner projects, one should have attained a level of comfort getting around in Scratch, how to work with code blocks, where to go, how to find what you need, dealing with multiple sprites and the stage, working in the three tabs, and even adding extensions. So what's left?

As we move toward competency, we also move toward independence. In many ways, the intermediate projects are about becoming confident and capable in making your own projects from scratch (from nothing). One of the biggest keys to opening the world of possibilities that Scratch provides is making your own art. The library is great to get started, but if you want the flexibility to take projects into any subject you want, you'll need new art. Getting comfortable and confident with all the power and potential of the Costumes tab is key to making anything you can imagine. Teachers need to get comfortable working with the art tools so that they can guide students through using them. When our students become comfortable and confident in making their own art, the possibilities become endless for them. They see the potential to follow their own dreams and can self-motivate to pursue their imaginations. Empowering creativity and self-fulfilment is the most powerful transformation we can achieve in teaching with Scratch.

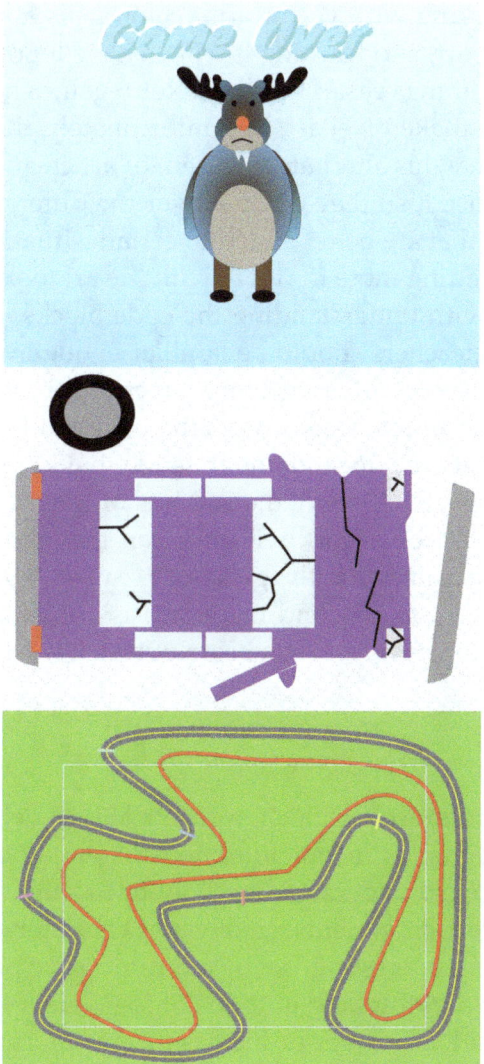

Figure 9.8 Three examples of the custom art created in the projects in this book. Learning to use the art tools effectively can help students (and teachers) take Scratch anywhere they want!

Learning the art tools will open up worlds of possibility, but you'll find that students often don't want to use the art tools in effective manners. They'll often default to just using the paint tool and crudely scribbling things in, then get frustrated when they can't make quite what they want, or struggle to correct it. Getting students to create vector shapes is a really important hurdle to clear. If you can get comfortable and confident using the Reshape tool, in turn you'll be able to teach your students to use it and open up a whole other level of creation. In my professional game work, I mostly focus on pixel art, and I have a lot of love for it, but it really is important we take the opportunity to get students working with vector art. Vector art is a fundamental part of professional digital art, and learning it will create a lot of possibilities, while also making a lot of things easier for them. Retro games have done a great job of inspiring kids to make pixel art, but unfortunately, that can work against us getting them to try this alternative system of art creation. It's worth pushing them to try to use it so they can discover the differences and learn new skills and methods that are better suited to certain situations.

In addition to getting more in-depth with the art tools, we should also be more comfortable with understanding the code blocks. At the intermediate level, students and teachers should be familiar enough with all the different code blocks. If you see a code block you haven't used yet, give it a go. You don't have to build entire projects around them, but you should be able to put them to use, see what they do, and understand what's going on. Ensuring we know the code blocks takes away the last bits of ignorance about our tools. Knowing what we have to work with, we can then focus on the methods, techniques, and most importantly, goals of creating. Everything is waiting for us, ready to be put to use, and the world is our canvas. We've finished our beginner apprenticeship and can begin our own work journeying toward mastery.

Logic

With our growing comfort in Scratch, we were able to delve deeper into logic structures and commands in our intermediate projects. Logic is the heart of computer science, so understanding how to use the ●Control blocks and ●Operators to create logic systems reliably and accurately opens up the possibilities of much more interesting, deep, and responsive projects.

Understanding and thinking in terms of logic are a great benefit for our students in any field. In an era of increasing misinformation and disinformation as well as political extremism, arming our students with strong logic skills will help them fight against the seemingly growing tide of fanaticism and suspicion. This is perhaps reason enough to teach coding, even if technology wasn't a dominant factor of industry and employment. We don't have to approach politics for this to be a benefit. Simply engaging students with logic, and making it a part of how they think, plan, and act through coding will help make logic a part of how they think outside of coding, and it will benefit them in so many of the facets of life.

Intermediate Check-In ◆ 149

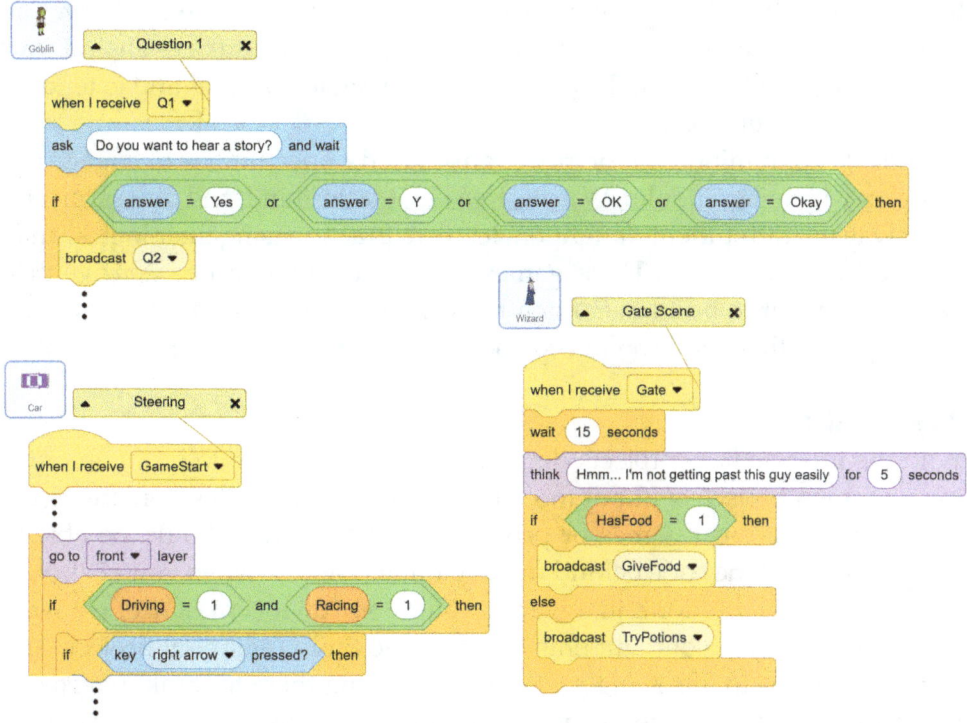

Figure 9.9 Some examples of how logic systems were used in the four projects in this book. Green Operator blocks help test values and states, and can be chained together using Or and And blocks for more complex conditions.

Specifically, though, we are dealing with coding in Scratch. How do we utilize these logic systems? How do we highlight them and reveal their utility and purpose? We've given a number of good examples in our intermediate projects, and there will be considerably more in our advanced projects. Thankfully, we can show students – and this is one of the great benefits of coding- and game-based education – real working models they can see in action, change, and test. When they miss a •<Not <condition>> block and things go wrong, they can discover the consequences as well as the complications and, through this, gain a true understanding.

An important skill to develop as an instructor, and to develop in our students, is the ability to walk through code. This entails reading code and following or predicting what it does. Practising reading code, especially whenever a bug shows up, is a great way to build our understanding of the underlying principles in code. To help students understand, we can try to create real-world examples of logic and code. If you walk into a room and it's dark, you flip on the light switch. In code this would be "When Backdrop Changes If Room is Dark = True Then Turn On Lights". It's an example I use a lot to describe If blocks only doing something when needed, by testing a condition. By building up a repertoire of examples like this, we can help make sense of abstract concepts outside of coding and Scratch and start building that realization of the principles applying to our lives and extending throughout our universe.

Also keep in mind a lot of great game-based opportunities to engage our students with logic. Sudoku, Wordle, Kenken, Mastermind, and others can be a great way to grab students' attention, let them practice using their logical analysis skills, and have fun. Those games tend to be a little more intensive than you might want to build in Scratch but can be great for students to attempt making when they move on to text coding with languages other than Scratch.

Project Design

With the intermediate projects done, you should feel able to jump on creating your own projects without too much worry. You've learned enough techniques to make most games or projects on Scratch. You should be able to imagine some of the code behind making things happen when you go browsing projects. But we don't just want to build our own confidence; we want to be able to provide our students the ability to build their own projects and bring their ideas to life. A big part of this is getting them to think in terms of planning and design. They need to take the techniques we've taught them and hypothesize and synthesize their own concepts.

A number of techniques can be very handy for thinking in design terms. Messages allow us to order all sprites and the stage to react simultaneously, creating distinct moments for events or scenes. We can use the backdrop to convey a change of scene or state. Custom blocks can create reusable code to handle common problems or processes. Menus can guide users to distinct opportunities or get input from them to customize their experience. Cloning can make temporary or duplicate objects simple. We want to have our students be able to think of how they'll achieve what they want in projects, imagining the possible solutions to the problem of creating what they want. Instead of just having them stumble toward their solutions in Scratch, competent intermediate students should be able to articulate what they want to make, and some of the solutions they'll use to make it happen.

Ideally, we move from students just trying things to students making plans, understanding interactions and events, recognizing their techniques, and being able to imagine structures and solutions. Unfortunately, it might be difficult for students to work together on the same project in Scratch, but we can still have them work together in project planning and design, as well as in testing, bug-fixing, review, and assessment. Try to think beyond just time at the keyboard and get students to make design documents, do mock-ups, and work in teams to plan projects and to review them.

10
Follow-Up: Extending the Projects

As we did in the first book in the series, the projects presented here are by no means the be-all, end-all of their concept. We developed them to be templates for you or your students to take and tweak, improve, and customize. We glossed over some areas for brevity and clarity you may wish to go back and add in. You may find yourself wanting to use different techniques for some things. This is great! We wanted you to take these projects, customize them, expand them, and make them your own. Intermediate Scratch coding is about taking things and remixing them, exploring all the possibilities and techniques. You need to be able to see ideas and make them your own – adding features, tweaking, removing, or replacing them. It's that code surgery that helps you master understanding not just code but how code fits together and projects as whole operate.

In this chapter we'll give you some more ideas about features you can add to the projects you've created. We'll go through each of the projects from this book and our previous book, *The Teacher's Guide to Scratch – Beginner* with ideas about how you can apply your newfound skills to make them better, expand them, and try out new coding concepts. You'll find these optional expansions aren't just a great way to build your skill set and comfort with coding but also provide you an excellent list of ideas to throw at advanced students that you want to keep busy and challenged. One of the best parts of challenges like these is seeing all the different ways they can be brought to life and interpreted, both in artistic terms and in coding techniques.

DOI: 10.4324/9781003399070-10

Commenting

When we go back to our old projects, we can find them harder to understand than we imagined because we've changed our own thoughts and perceptions as we've grown. If we remember to think of coding as language, we can perhaps understand the change in more human terms. Who as an adult still speaks like they did when they were a teenager? We learn new phrases, get new catchphrases, settle into new habits, accents, or mannerisms.

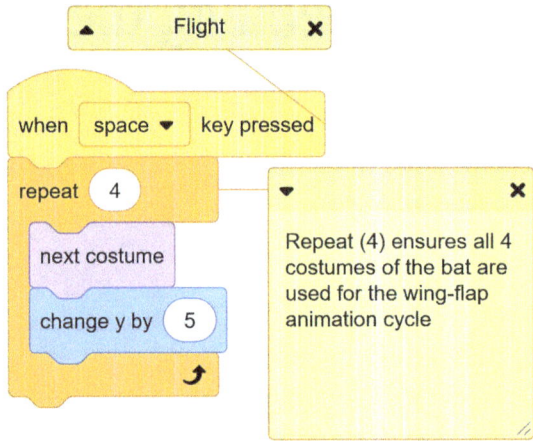

Figure 10.1 Comments can be collapsed (Flight) or expanded. When expanded they can be resized to any shape by dragging the bottom right corner. Switch their state by clicking on the arrow in the top left corner, or delete them by clicking the top right "x".

Because coding requires us to both translate and interpret as well as keep a mental model of the project or processes, it can be very difficult to read other peoples' code, and our own code if it's been too long since we last laid eyes on it. One habit we should get into, along with our students, is commenting. Commenting is the practice of providing text hints and descriptions added to a project to explain things using natural human language. Comments in Scratch take the form of sticky notes, either at large or attached to specific code blocks. Simply right-click to add a comment on the background for an at-large comment, or on a code block for an attached comment. You can resize comment notes as needed by clicking and dragging their bottom right corner. You can see I used comments to label all the stacks in my projects, but this is just one minor use. Good practice would be to be more descriptive and especially mention any connections with other code or objects.

Try to go back through the projects and add comments to describe what stacks do, what variables are for, what objects will react to what broadcasts, etc. Add notes to your projects that let you read and easily understand its structure, flow, and processes at a glance. This is a key skill for professional programmers and helps immensely when two or more people are handling the same project. Getting used to doing it will greatly improve students' skills while making evaluating their work easier as well.

Figure 10.2 The four projects covered in our first book in the series: The Teacher's Guide to Scratch – Beginners: Professional Development for Coding Education.

Dinosaur Dance Party

As our first project, our Dinosaur Dance Party was pretty simple. It did provide a good basis on how to work with animations and music in Scratch, but what more can we do? Now that you've got more experience and techniques under your belt, let's try to add some more features: Can you allow the music to be switched by the user? What about some pyrotechnic effects? How about having the dinosaurs sing out some lyrics? Can you add some custom vocal tracks? How about some lyric subtitle displays so users could sing along to a song?

Fireworks Display

In our Fireworks Display project, we had the basics established to plan your own fireworks shows. Hopefully, you took the opportunity to plan out a little sequence of your own with different firework designs. Now we can look back on it with our newfound skills and consider some changes or additions: Could you switch the whole system over to using clones so launchers launch any number of fireworks anytime they're called? We can also look beyond just the fireworks for ideas – what about the backdrop and setting? Can you build a scene for your national capitol? What about having the fireworks spawning some puffs of smoke that hang in the air before disappearing?

Batty Flaps

Flappy Bird was so successful because of its simplicity, so it can be a little hard to add to. You could add different types of obstacles – ones with only the top side, bottom side, or the full gates we have now – and determine a way to switch between them. In the broader genre of game, the "Endless Runner" can offer plenty of other ideas and similar games to try to build. Perhaps make a dinosaur-jumping-cactus game instead. Or try a multiple-track game where we choose set tracks or rails to jump between to bypass the obstacles or collect things.

Butterfly Catcher

Our Butterfly Catcher game had some excellent built-in difficulty scaling, so it can inherently match the user's skill level. But what if we added a twist? You could try adding in movement patterns that are more than just step and bounce. You could even add in another layer of difficulty with having temporary vision blocking obstacles like falling leaves. For looks, maybe you could improve the butterfly's animation and try to slow it down or smooth it out – without affecting gameplay.

Figure 10.3 The four intermediate projects we cover in this book, the second in The Teacher's Guide to Scratch series.

Pen Tool Fun

In our Pen Tool Fun project, we were pretty exhaustive in drawing basic shapes, but there's plenty more to discover. Can you figure out how to draw a spiral? What about a classic five-line star? Can you add in the ability for the user to move the pen without drawing other than dragging with the mouse? Can you draw a grid? How about allowing the user to control the size of drawings? What about letting the user freely draw with the mouse like a digital pen?

Interactive Story

This project was developed very much as a template, showing the framework and technique to make your own interactive stories. There are still lots we can do with it. By adding a menu, perhaps you could get your basic info for the game instead of having the goblin intro like in the original project. You could ensure every scene has a click-to-advance design instead of the numerous scenes that automatically advance. A mouseover system to give little pop-ups of information about things without players clicking them could be a great way to add more context or warning for players to help them explore the story world. You could add in vocal tracks or the text-to-speech extension to narrate or voice-act the story to make it more accessible for the visually impaired.

Snowball Fight

Artillery games like our Snowball Fight traditionally have a lot of options for gameplay, so there's plenty we could add to this game. You could explore any number of different "weapon types", like a special triple-snowball attack, or a sloppy slush ball, or a melt ball that could reduce the wall obstruction. Adding a shot-tracing feature to give a reminder of previous shots trajectories could assist players and help highlight the physics in the game. The wind could vary within a bout rather than only between bouts. Likewise, you could have the target potentially move during a bout. You could allow the player to move to some degree so they can get different trajectories.

Big Map Racing

This project is perhaps one of the most technically challenging with its Scratch rule-bending design. One can relatively easily include a number of road hazards to add in more dangers to the course; crashed cars, tires, or tools could all litter the raceway for added challenge. Can you figure out how to have them generate randomly, only on the road? An oil slick obstacle or wet track hazard could be another interesting addition influencing steering rather than straight up crashing the player. You could add speed boost pads to provide ideal paths. You might add in more terrain types rather than just the simple on/off-road condition that could affect speed or steering differently.

Advanced Scratch

We aren't finished with expanding and reimagining these projects. In *The Teacher's Guide to Scratch – Advanced*, we'll revisit all 12 projects in the series with ideas on how to incorporate advanced Scratch techniques. We'll have ideas that will both push your coding and your creativity to new heights with new methods and insights to give you some fun and interesting challenges for yourself and your students. So be sure to check out Book 3 for our most ambitious and interesting coding challenges yet!

11

Troubleshooting Scratch

Perhaps nothing strikes fear into teachers told to integrate coding into their classroom more than the thought of dealing with bugs, errors, and computer trouble. Admittedly, coding can throw a lot of surprises at you. Earlier, we even said bugs are a part of the process. This might not be very reassuring talk, but just like coding itself, we can prepare ourselves for these eventualities. Here's some advice for the most common problems faced in the classroom when teaching with Scratch, in an attempt to arm you with the knowledge and practice to overcome most of the potential issues you'll face.

Demonstrating and working through bugs and errors with students is a fundamental part of working with technology. We want to show them resilience, determination, logic, and best practices for dealing with setbacks, and that's true with any subject, be it digital or analogue. By taking the time to make bugs and errors part of our teaching, we can use them to touch on a lot of key skills and make opportunity out of crisis. Working through bugs can build confidence and resilience, teach solution-finding practices, help practice analytical thinking, and provide for social learning opportunities with students helping each other with brainstorming, analysis, planning, and helpful solutions.

Site Issues

If you're using the website version of Scratch (as opposed to the offline downloaded and installed program version of Scratch), there can be some connectivity, browser, or website issues. That's without considering all the computer and connection issues – computers not booting, students unable to log in to their profiles, forgotten passwords, computers deciding now's the time to install updates, Wi-Fi not connecting, etc., which are all local issues with machines and networks. We are not dealing with those issues because they'll be specific to your school and school board, so hopefully you've been through all those issues with your local tech support. Instead, let's focus on issues particular to Scratch, assuming you've successfully reached the website.

Problem: The project is freezing/not responsive.
Possible Solutions:

1. Refresh the website (click the Refresh button in the browser).
2. Does the project use clones? If so, check if too many clones are being generated (over 300 can overload the website). Clones, repeats, or My Blocks generating clones is a common cause of this issue.
3. Close and restart the browser.
4. Reset the Internet connection.
5. Give it time (website may be temporarily down or overloaded).

Problem: Can't remix projects.
Solution: Log in to a Scratch account.

Problem: Can't autosave or save.
Solution: Log in to a Scratch account.

Problem: Can't name projects.
Solution: Log in to a Scratch account.

Problem: Can't access/save to/see the Backpack.
Solution: Log in to a Scratch account.

Problem: Can't access previous projects.
Solution: Log in to a Scratch account.

Problem: Can't find/see the project that was sent.
Solution: Project owner must ensure the project is shared.

Problem: Link doesn't work.
Solution: Project owner must ensure the project is shared.

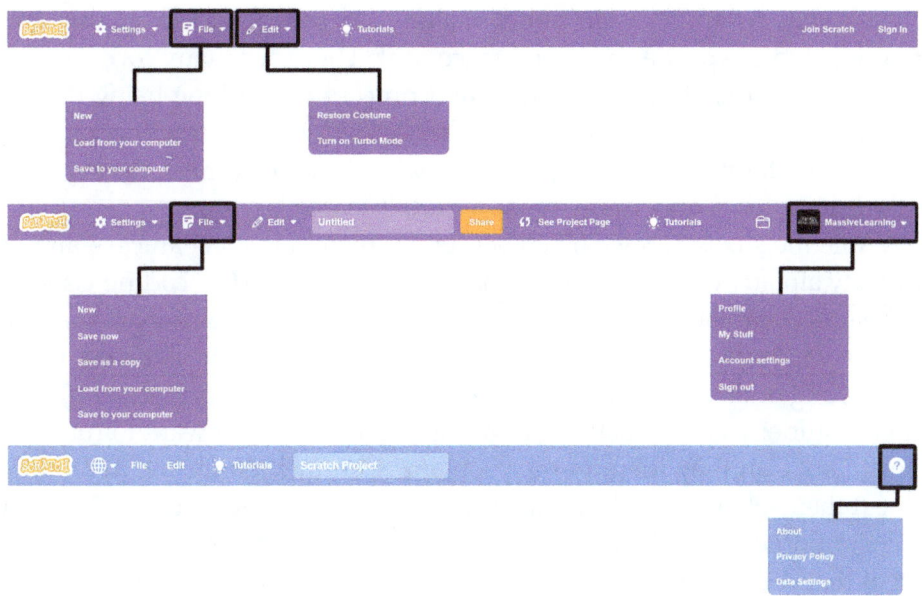

Figure 11.1 The Status Bar's File options when not logged in (a guest), when logged in, or in offline Scratch. Cloud saves and sharing are not available unless a user is logged in to the website!

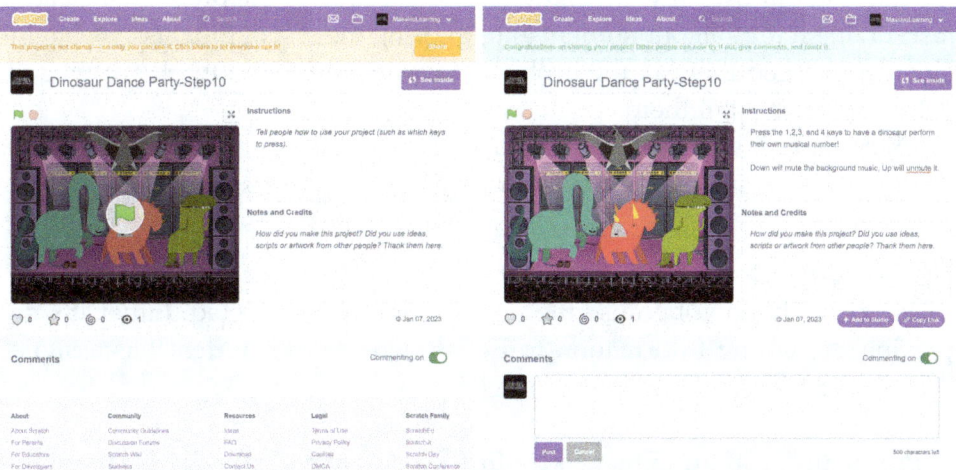

Figure 11.2 You don't have to share what you create, but it does help build the community. For education it can allow teachers to share starter projects, or for students to do peer review or submit projects. Here's a project before and after sharing.

Coding Issues

Most errors and issues you face will be coding problems. Computers are very literal in their understanding, and this is something most humans are not very good at. Our brains are built to predict, imagine, and leap to conclusions to act quickly and save time and energy. We often make assumptions or leaps that computers can't and, making our code, end up doing something we don't expect as a consequence of it. There are many forms of error and many reasons, but the fundamental process of thinking being different between humans and computers will likely always cause friction and bugs. The most irritating truth of bugs is, no matter how unwanted the behaviour is, the computer is only doing what we tell it to do.

While the site issues we listed earlier are relatively mechanical – and you probably already deal with those kinds of issues regularly – coding issues are a whole new can of worms. Most importantly, we want to prepare ourselves for how we think and deal with bugs and errors before we even start fixing them. A bug or error is, like in any other subject, an opportunity for reflection. When a student says something is going wrong, we don't want to rush to fix it for them. Can they identify the problem? Can they enunciate it? Can they solve it themselves? Can they eliminate some of the possible causes? Our first goal should be to challenge the student to overcome the obstacle, or at least give it their best shot. If they engage and try those things, we still don't have to leap in ourselves to fix it. We can then challenge the class to think about the problem, discuss it, and suggest solutions. This helps take the problem and turn it into opportunity for practice and class engagement. If that fails, then we can lead our own review, analysis, and solution to the problem instead of just solving it for that one student and ignoring the rest of the class. By having already engaged them in the problem, our time spent fixing the problem is helping inform all of them.

Now that we have some better idea of how to approach bugs, what are we dealing with? There are four basic areas where things go wrong.

The Wrong Object

These errors happen because the wrong object was selected. Either it's displaying the wrong information to the student or the student has assigned code or properties to the wrong object.

> **Problem:** "I can't find that code block" or "That code block isn't there" – the list of code blocks is different than expected.
> **Solution:** The student has selected the stage instead of a sprite. The stage doesn't have a number of the code blocks available to it since it can't move or do a number of other things that sprites can. Make sure the student selects the correct sprite first.
>
> **Problem:** The wrong object acts or reacts.
> **Solution:** Code was placed in the wrong object. Check the object that did act/react and the one that didn't, and you'll likely see some code assigned to the "wrong" object.

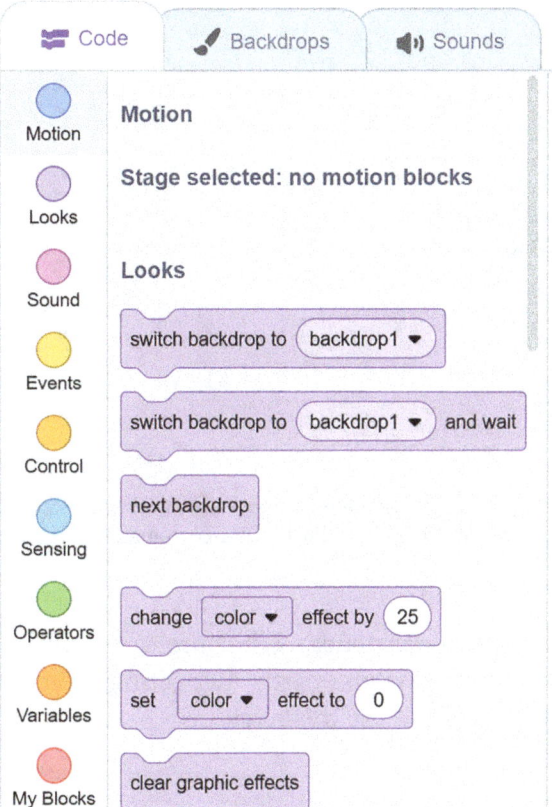

Figure 11.3 When the stage is selected code blocks that don't apply (such as all Motion code blocks) don't appear

Alternate Solution: The wrong broadcast event might have been called, making an unexpected reaction happen. If the behaviour you saw is supposed to happen at a different time, that would indicate a wrong event instead of a wrong object issue.

Alternate Solution: When using a ●((Property) of (Object)) driven behaviour like ●[Move to X: (●(X Position) of (ObjectA)) Y: (0)], if you selected the wrong object, you could be referencing the wrong object's properties and making it look like one object is calling the behaviour rather than the other. Double-check any ●((Property) of (Object)) code blocks and see if they are correct.

General Tips

Transferring code blocks. Remember all the ways to copy and transfer code blocks (drag, copy + paste, the Backpack). You don't have to rebuild things from zero. Save time by copying code that's incorrectly in one object over to the one that actually needs it to save time (but remember to delete the incorrect code after copying!).

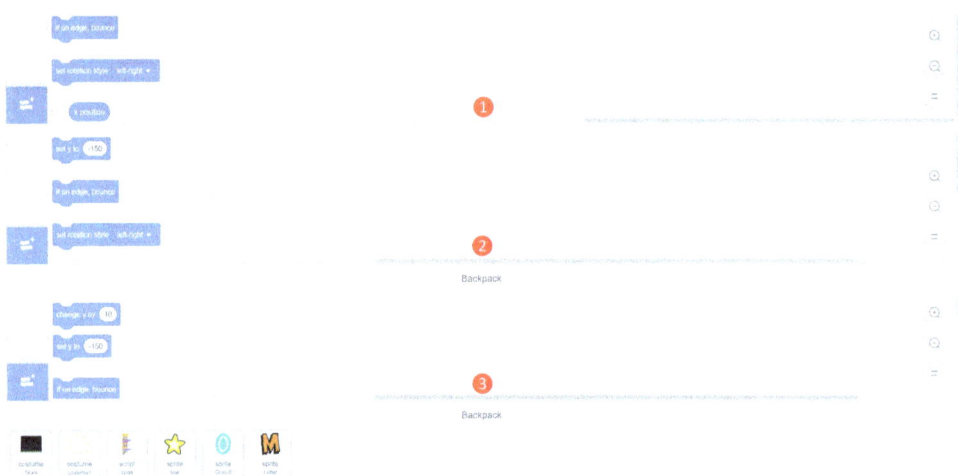

Figure 11.4 You must be logged in to access the backpack. ① Shows none. ② Shows the minimized backpack. ③ Shows the backpack maximized with some costumes, scripts and sprites already added.

Specify Objects. While instructing, be sure to specify what object you're coding for and when you switch objects. Repeating yourself to the class can be very useful for avoiding these mistakes.

The Wrong Block

There are a number of blocks whose similar nature can easily trip up new coders. These errors can also be hard for people to spot because of our habit to read what we expect rather than what's there. Knowing the likely culprits, you can look for them easier and quicker, as well as predicting the issue and making sure you clarify the difference when they come up in instructed learning.

Confused Pair: Left vs. Right

The turn code blocks for left and right turns can easily be confused, especially for younger students still uncertain with their directions. It's unlikely to cause serious issues, but if something is veering off in an unexpected direction and you used a right- or left-turn code block, check if there was a switch. Sometimes, seeing the difference can actually be fun and interesting, like in our Pen Tool Fun project.

Confused Pair: Go To vs. Glide To

Both of these code blocks can send a sprite to a specific X/Y or sprite location, but they work differently. •**[Go To]** automatically teleports a sprite instantly, changing its X and Y properties. •**[Glide]** works by changing the X and Y properties over a set amount of time. Kids love glide because it animates, and

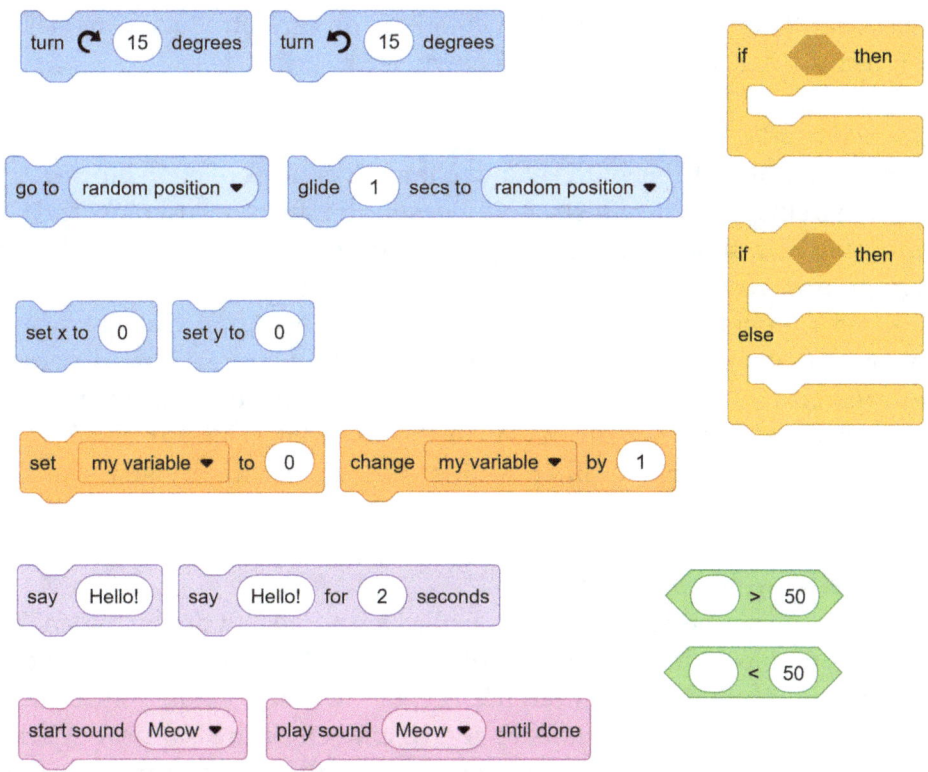

Figure 11.5 The most likely code blocks students will mix up. If your project uses one of these, expect somebody to grab its twin!

they can see the sprite moving (gliding) to the destination, but it can cause some issues if used in place of ●[**Go To**]. Glide includes a time value, and this acts as a hold or delay on the code. Until the glide is complete, the computer won't move past that code block, which can cause some issues. If you want a sprite to move with glide and to detect collisions, you can't use both in the same stack, because while the glide is happening, it won't run the collision detection code because it hangs on that single timed code block. ●[**Go To**] can cause its own problems, as it changes instantly, so sprites look like they're teleporting and might be too sudden a change; you could even teleport past an obstacle without ever colliding with it like you expected.

Confused Pair: X vs. Y

It can take a while for X- and Y-coordinates to sink in. Even adults who have learned them can forget when they haven't been using them. When things move left/right or up/down unexpectedly, you can always double-check any X- or Y-affecting or powered code blocks. In the Properties panel, the X and Y properties have appropriate arrows beside them as a reminder of which is which, but there's also plenty of mnemonics you can introduce your class to (X is a cross/X is across, or Y points down, etc.).

Confused Pair: Set vs. Change
There's a lot of features with set and change versions of code blocks: X/Y positions, size, graphic effects, volume, sound effects, variables, and pen properties. Until students get used to working with variables, it's very easy getting confused between the two. Set makes a property/variable an exact number, regardless of what it was (absolute). Change adjusts a property/variable from its current position (relative). This difference between absolute and relative changes can be hard to describe, but projects can be a great way to visualize the difference, and it can be well worth taking the time to illustrate the difference with a visual example like position or size.

Confused Pair: Say/Think vs. Say/Think For
All four of these code blocks allow displaying some text on the screen. The **[Say]** code blocks use a speech bubble, while the **[Think]** code blocks use a thought bubble. The difference is "For", where a value is introduced, allowing a time scale that makes the displayed text disappear when it is up. Without "For", the text is permanent and only disappears with a **[Hide]** or new **[Say]**/**[Think]** code block. The permanent versions are good for non-action, non-animating scenes, information displays rather than reactions or conversations. One needs to be more involved in ensuring they disappear when you need them to. "For" versions are great for fast-paced programs, where you want to maintain some action and activity, and you can set and forget them since they'll disappear on their own. Be careful the information displayed doesn't disappear too quickly for slower-reading users.

Confused Pair: Play Sound vs. Start Sound
Similar to **[Go To]** vs **[Glide]**, the difference between these two blocks is time. A **[Play (*sound*) Until Done]** code block plays a sound effect, but it holds the program on that code block until the sound effect has finished playing. A **[Start Sound (*sound*)]** code block begins playing a sound effect but doesn't hold the program (other things can happen while the sound is playing), and code execution continues on without any delay. So **[Play (*sound*) Until Done]** is useful for things like alerts or dialogue, where you want to pause things until it completes, whereas **[Start Sound (*sound*)]** is great for sound effects and other incidental sounds that may even stack up and simultaneously play while other things happen.

Confused Pair: If vs. If/Else

This pair confusion won't tend to last long with the block structures limiting coding options, but I thought it is worth mentioning. It can be hard for younger kids to see an •[If <True> Then] code block and realize it isn't the one they want, so you need to be very clear about selecting the right one with younger audiences. The difference is, of course, the "Else" clause. An •[If] only tests a condition and allows code if the condition is found true. In an •[If/Else], the condition test determines which of the two clauses of contained code is used, the first if the condition is true, the second if not.

Confused Pair: > vs. <

This will be no surprise to teachers, coding experience or not. The •Operators blocks for numeric comparison can be confusing to younger students. Again, the key is clear communication beforehand to ensure the correct selection in the first place. We can use all the math class mnemonics to try to get the difference clear with our students. The nice thing about code is that with the computer responding immediately to the code, students can test and see the results on their own.

The Wrong Order

Even with the right sprite and the right code blocks, things can go wrong. Our third category of errors are "wrong order" issues, where code blocks aren't in the right sequence to execute as desired. This is where we start getting into more complicated issues. Thanks to Scratch's code block system with distinct shapes and colours for code blocks, a number of issues are much more visible and easily avoided. This visual feedback doesn't just help students follow and build things but can also help avoid or correct bugs.

Simple Sequences Code Flow

The most basic errors in this category are simply putting a code block above or below where it belongs. In many cases, the exact order doesn't matter. For example, in giving multiple variable values to initiate a program or setting multiple properties to represent a new state. In these cases, the exact order rarely matters; the properties are all set in a single frame of animation and don't interfere or influence each other. But in other cases, the order is extremely important. Take for example a sprite given a •[Move (100) Steps] then •[Turn Right (45) Degrees] versus a sprite given a •[Turn Right (45) Degrees] then •[Move (100) Steps] instructions. They'll end up facing the same direction, but almost as far from each other as they are from their starting positions.

Control Structures and Code Flow

As we introduce control structures like •**If** statements or •**Broadcast** events, we change how the code flows. Thinking of code through flow charts or process systems, errors can occur because we took a wrong turn in our chart. This could happen because we called the wrong •**Broadcast** event or were listening to the wrong one, triggering the wrong code to execute out of sequence. It could occur because we changed a control like a game state that triggered an •**If** statement at the wrong time and changed the expected sequence of events or a property at the wrong time. •Control structures are very powerful and handy, but a "typo" of grabbing the wrong block or value can have us swerve off course in the flow of processes.

Logic Clauses and Chains

Learning to use logic can be a challenging task; there are lots of errors to be made when including or excluding conditions in code. It can be hard to know whether to write some conditions inclusive or exclusive, whether to use an "If", or an "If/Else", or which two clauses to put your code in. If something strange is happening, always check the logic conditions; it's surprisingly easy to end up with the reverse of what you intended because of our love of leaping to conclusions, leaving the cold mechanical reasoning of computers in the dust. One very easy issue to fall into is chaining logic conditions – when you put multiple logic operators together for a single test. <<<A>or <B>> and <C>> will give different results than <<A> or <<B> and <C>>>. It can be very hard to spot this kind of error, so know to look for it!

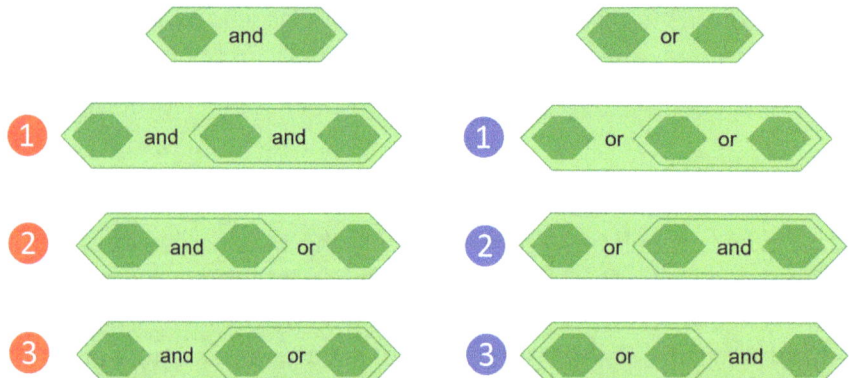

Figure 11.6 Combining logic blocks can easily become confusing. It's important to remember how they are connected may affect the results. ❶ Ands can be combined and simply add to the necessary conditions to meet. ❶ Ors can be added to add additional options to meet. When mixed the outer code block is dominant. So in ❷ you must have both the first conditions or have the third. In ❸ you must have the first and one of either of the second and third conditions. In ❷ you can have the first condition or both of the second and third. In ❸ you must have one of the first or second conditions plus the third.

Concurrency and Race Conditions

We can run into other problems in the flow of our code processes separate from those previously mentioned. Sometimes issues happen when we need to handle multiple processes that happen concurrently, or with (near) simultaneous execution of code. This can cause specific problems with order of execution. If any properties or variables are affected in one process that affects the other, the order of operations could drastically change the results. It can be challenging to make simultaneous systems work without improperly interfering with each other. When we don't know which process will complete first and they can influence each other, we call that a "race condition". The faster process "wins" and makes changes that could affect the other. Sometimes, the order can be predicted, sometimes not, so these issues can be very tricky. We can try to put code into single stacks or split it into multiple stacks to simplify processes to a linear run or to ensure concurrent execution as needed. Single stacks make the flow much easier to analyze and predict, but sometimes concurrency is exactly what you need. As we mentioned earlier, take for example the •[Glide] code blocks: if an object is gliding and you need it to be clicked or collided with, you'll need to put the collision or clicking code in a concurrent process because other code in that stack won't execute until the glide is completed. Delay blocks may or may not help simplify these issues. Concurrency can be a blessing and a curse and usually very hard for the human brain to predict accurately. If something isn't happening, or it's happening at the wrong time, check for any block with a time value or time-sensitive word like "until", "for", or "done" and think about how they might be holding things up.

Other Errors

These three categories merely represent the most common issues a coding educator has to deal with. In this category, we'll deal with some of the other less-expected issues that can happen in Scratch that don't necessarily fit into a convenient overarching category.

Layers

Some of the simplest issues come from layering. Every graphic on the screen is printed in layers, each object laid on top of the objects below it, all on top of the backdrop. At times, this layering may cause issues. Large objects may completely obscure smaller objects underneath them. You may need to control the depth of objects at times, but the •[Go To [Front]], •[Go To [Back]], •[Go [Back] (1) Layer], and •[Go [Forward] (1) Layer] code blocks can be used to solve these issues.

Visibility

There are a number of issues you may run into dealing with object visibility. The first and most obvious is having objects disappearing. If you ever •[Hide] an object, you will need to have a •[Show] code block somewhere to make it visible, or you'll never see it again after it first disappears. While invisible, or hidden, a sprite has limited functionality. A hidden object cannot •[Say] or •[Think]. A hidden object also doesn't collide with other sprites.

Colour Selection

Scratch uses a graphical technique called anti-aliasing. This is a process that blends (and changes) colours to smooth out lines to appear more visually appealing and avoid the jaggedness that pixel designs and displays can cause. This looks great, but it can lead to a few tricky issues. If you are using any colour collisions, watch out for anti-aliasing blending the colours you think are involved. If you want to use a colour and use the colour picker, you may see that there are many more colours than you expected in the zoomed view of the screen. Any colour near a line or change of colours may end up being blended by anti-aliasing. This means that colour collisions looking for that colour won't find it because it changed to a different colour. This is especially likely to happen with sprites that are being displayed at less than 100% size or at anything but 0% ghost effect, which blends them with whatever is behind them to create the transparency effect. Colour collisions are great, but if you're having unexpected results, it may be because of these issues.

Clones vs. Originals

As we mentioned in our projects, clones can be tricky. Whenever you're using clones, remember that there's a difference between the original sprite and any clones you make. The original sprite will never receive the •[When I Start As A Clone] event, and the clones will never receive the •[When ▷ Clicked] event. Don't expect the original to act like a clone, and don't expect clones to act like the original. It can be hard to get used to it, but once you do, it is a very powerful system to work with. Also, remember that clones always start with the same properties as the original. If you don't see clones, is it because you had the original •[Hide]? Unless you •[Show] the clones, they'll remain hidden.

Wrong Concepts

The hardest of bugs is when we simply had the wrong conception of a system or method. Our brains ran off with an idea that wasn't true, doesn't work that way, or otherwise doesn't line up with reality. There's no predicting this and no preventing it. Sometimes we have ideas that don't work. We want to try to make them work and account for any bugs in the system, but sometimes the way we conceived something is simply not how it will actually work. Maybe we didn't account for some factor that makes it unable to work, or it is simply inefficient or impractical. It happens. Even to the pros. The way to address this issue is always approaching coding with a flexible mindset. There are many ways to address an issue, and sometimes we need to explore and try some options in order to really understand the problem or our tools in order to come up with better ideas. Learning can sometimes mean failing. Learning that something doesn't work, or work in that way, or work in this condition, is still learning. Learning those limitations and finding ways to work around them is perhaps the greatest part of learning coding.

Backup Plans

Having been a coding educator for years, I've seen just about everything go wrong. The worst issues are large-scale technological issues out of your control: the Internet is not working, power outages, websites go down, etc. Most of the time, educators can simply switch tracks and work on something not so technologically dependent, but for those of us that are coding specialists, how can we deal with the wrenches that sometimes fall into the gears? Here are some ideas about how to work around dealing with major technological hurdles and still deliver some educational opportunities.

Offline Scratch

One of the most impressive things about Scratch is that they have both the online instant access website and a downloadable offline version. If you have issues with spotty or slow connections, you can download it and install as an app that requires no Internet connection to work. This can be a wonderful tool for rural and remote communities, to ensure lack of Internet doesn't lead to lack of learning access.

Some educators may actually prefer the offline version of Scratch. By having the app, students won't be distracted by the shared content on the Scratch website. While I think the inspirational factor of the sharing platform side of Scratch is a great benefit to students, some may benefit from less distractions.

Another factor going for the offline version of Scratch is that it means students will save and sort their projects in their student profiles. This does mean the classroom organizing and viewing options aren't available but for prolific creators, but the ability to sort projects into folders in a way that personal accounts don't allow for can be a big benefit as dozens or even hundreds of projects pile up.

Most school board computers will have program installation locked down (and for good reason), so if you want to work with offline Scratch, you'll need to have the tech department install it on the student computers.

The nice thing is, this can mean having someone else do the work and possibly automate the process, so it shouldn't be any additional hassle for teachers to have this capability in their classroom.

Pseudo-Coding

Let's be clear: "pen-and-paper" coding is absolutely no replacement for working with computers, but learning to do pseudo-coding can be a powerful and useful skill to add to the mix of tech skills we develop in our students. Pseudo-coding is when a coder writes out their design, not using the exact code word for word, but in a shorthand to give the broad strokes of their plan on how to achieve or organize things. It's more about planning than syntax and details.

Pseudo-coding can look different for different people, as each person makes plans or notes in their own way. In general, pseudo-code will help define the components of a project – such as the objects, backdrops, costumes, sounds, writing needed, as well as the processes, the events, and the stacks that will make things happen. Pseudo-coding can be as verbose or as brief as needed; plans can be quick sketches or fleshed out fully to the point of actual code, but often it consists of shorthand sentences, like "If near [objective] then change to warning state". Maybe they don't know what they want the warning state to be, or maybe they do, or maybe they have some idea about a feature like "Repeat 10 {size +10%, wait, size -10%}" as a note to make something pulse and a method to achieve it. You'll note the exact name of code blocks might not be listed, but the intent is clear.

In addition to more written pseudo-code, there can be very different methods for keeping these notes, as the creative process works differently for different people. Some may use more text document format, others may want to use spreadsheets to list ideas, some may prefer physical whiteboards, others might use presentation software to make things more visual, and flow charts could be used, mind maps, or even wikis for very large projects. No perfect system exists for pseudo-coding, creative designing, or visioning, since everyone and every project is different. Explore and experiment with different methods and tools, see what you like, and help your students explore all the ways their ideas can be brought to life. There are so many wonderful ways to be creative, and such wonderful things to create. You and your students have beautiful, thoughtful, meaningful, and fun treasures hidden inside them; by discovering and sharing the tools and opportunities to bring them to light, we all benefit. We hope this book has helped you learn, grow, and prepare to be a guide on that process and help usher in a beautiful era for human creativity and potential.

12

The Next Step in Your Coding Journey

You've now worked through, and are hopefully comfortable with, our intermediate Scratch training. Ideally, you'll have also worked through our beginner Scratch training in *The Teacher's Guide to Scratch – Beginner* to make the most of our challenges. At this point, we hope you've not only become comfortable with Scratch and coding but also confident in approaching new projects and have a good sense of what Scratch can do.

Figure 12.1 The three books in The Teacher's Guide To Scratch series.

DOI: 10.4324/9781003399070-12

The projects in this book had a lot more objects than in our first book; as well, projects extensively used state machines (game states) to determine behaviour. This increased the complexity of projects that can make beginners a little confused, but having become familiar with the basic functions and blocks, we could now look to put them into more complex structures and relationships. We used this new level of project complexity to explore a lot of new ideas. We had projects that were games, drawing systems, and storytelling platforms. We learned about the Scratch extensions, started using some simple physics modelling, and learned some very powerful techniques to increase the space we had to work with, with scene changes, level changes, and oversizing techniques.

So what's left?

In our third book of the series, *The Teacher's Guide to Scratch – Advanced*, we'll be taking things to the next level. We'll have another four projects that will introduce new concepts and features that will help you expand on your already-formidable skills. Book 3 is about taking your confidence toward mastery. We'll have big, complex projects that push the boundaries of Scratch. We'll start working more on professional skills and thinking. We'll have projects that are object-rich, and objects that track both global (project-wide) and local (just their own) states. We'll work with more custom blocks to see more of what they can do. We'll also use lists and see how we can work with data sets, including importing and exporting data files in Scratch.

Figure 12.2 The four advanced projects we cover in the series third book: The Teacher's Guide to Scratch – Advanced: Professional Development for Coding Education.

Our advanced Scratch projects would suit grades 8 to 10 and can offer more opportunities for storytelling, game creation, data processing, and professional skill development. They are a good introduction to working with Scratch at the top levels of what the platform can do, and a good springboard for moving onward to text-based coding and professional languages and coding platforms.

I hope you've had a fun time getting comfortable with Scratch and exploring its many uses, and I hope you'll join me with Book 3 to take your skills to the next level!

13
Final Thoughts

You've now got intermediate-level Scratch under your belt. Expanding your knowledge of the basics with these four projects, you should have a good idea of what Scratch can do and, even better, how to do it. When you bring intermediate-level Scratch into the classroom and get your students comfortable at this level, you'll have a crew of digital creators and explorers. They'll have the fundamentals down so they can pursue their own ideas and try almost anything with Scratch. They won't always succeed, and they'll need you for guidance, reassurance, pep talks, and challenges, but you'll have a whole new tool to explore any concept or topic you want to cover, one that has the students drawn in, engaged, and exploring with whatever content or curriculum you need to cover.

After completing these projects, you should be able to muster the courage to take on the challenge ahead of you. It can be daunting, it can be scary, but it will be a tremendously creative and rewarding experience. Our students need to see leadership and daring, but they can also see vulnerability and curiosity in the face of ignorance. We are role models, but we are not perfect or unstumpable. Embrace the errors and uncertainty we face in teaching coding. As long as we are technically proficient, we'll be able to lead or assist our students in thought experiments to find solutions. These challenges and the effort and methods to overcome them are perhaps the greatest lesson we can teach them, regardless of subject.

With intermediate Scratch, we've got ourselves and our students thinking in processes. Logic and flow should become ingrained, helping power students' critical thinking for any subject. You will have opened the door to the deep fields of computer science and information theory yet be able to keep things fun and engaging. This is the power of coding, and especially of game development, in education. If we can pair this deep technical knowledge and capacity with ethics, philosophy, and the thoughtfulness of the humanities, we will have given our students the greatest tools we can to understand the universe and build a brighter future.

Glossary

▷ – This icon is used to represent the green flag that is used as a start button for Scratch projects. It can refer to the start button in the Scratch editor above the Stage Window, or it can by in a •[When ▷ Clicked], which is the event that runs when a user clicks the green flag button.

✎ – This is the icon we use to represent the Pen tool extension in Scratch. The Pen extension, when added to a project, adds an additional category to the Code Block Library; instead of a coloured dot like other categories, it has a pen icon. When you see this icon, it means the following code block is in the Pen category.

.sb3 – This is the file extension for Scratch 3 saved files. Earlier versions of Scratch used .sb1 and .sb2. You can *"save to your computer"* and download an .sb3 file of a project and *"upload from your computer"* to load a file into Scratch.

Abstraction – This is one of many terms used to define computational thinking. It is the process of breaking a problem or goal into simplified and standardized components so simple and reusable processes can be developed to address those components. See also *computational thinking*.

Algorithm – An *algorithm* is simply a defined process to find a solution or achieve a goal. Math formulae are examples of simple algorithms. Recipes are another form of algorithm – a clear set of delineated instructions leading to a goal: food! Computer programs may be an algorithm or contain one or more algorithms.

Algorithm Building – This is one of many terms used to define *computational thinking*. It involves creating defined processes that, if followed, solve, or help solve, a portion of a defined problem or work toward a particular goal.

Argument – In computing, an *argument* is a piece of data passed to a function. Arguments are used by the function to evaluate a procedure or subroutine. They may also be referred to as *values* or *parameters*, though each has its own similar but distinct specific definition.

Art Workspace – In Scratch, it's the central portion of the editor when in the Costumes tab or Backdrops tab. It is comprised of the art tools and the canvas.

Asset – In general computing, an *asset* is something, not code, that a program uses to complete its function. This commonly includes graphics, sound files, 3D models, or data files.

Avatar – A person's representation on a digital platform. An icon, sprite, or even 3D model may serve as an avatar, depending on the context.

Backdrop – A graphic displayed by the stage in Scratch. The stage can have multiple backdrops, but only one can be displayed at any given time.

Backdrop List – When viewing the Backdrop tab in Scratch, the Backdrop List is displayed on the left-hand side of the screen. It shows all the current backdrops assigned to the stage that could be displayed and highlights the one currently selected.

Backdrop Tab – In Scratch, when the stage is selected, the second tab in the upper left-hand corner of the editor becomes the Backdrop tab. If clicked on, it provides access to an art workspace to create or edit backdrops and shows the backdrop list on the left hand side.

Backpack – In Scratch, if a user logs into their Scratch account, they can access the Backpack at the bottom of the editor. A small button labelled "Backpack" will show when collapsed, and clicking it expands to show a single row of thumbnails of what is currently in the Backpack. Users can drag scripts, sprites, backdrops, or costumes to the Backpack, where they will be saved and be able to be copied out from into any project the user opens. Items can be added to the Backpack by clicking and dragging them onto it. They can be deleted by right-clicking on them and choosing Delete on the pop-up.

Boolean Block – In Scratch, these code blocks are distinguished by their sharp, poky sides. They are used to provide arguments for some ●Control blocks, such as ●[If <condition> Then]. They are used to perform an evaluation that will return true or false, also known as a Boolean result. Most Boolean blocks are in the ●Operators category, but there are also Sensing category Boolean blocks and others. The ●<(0)=(0)> code block is an example of a Boolean block.

Broadcast – In Scratch, *broadcasts* can refer to a few different ●Event code blocks or to the unique named signals called Messages that are broadcast or received through them. Broadcasts are triggers that can allow code to trigger other code, including across multiple sprites or the stage.

C Block – In Scratch, numerous code blocks are shaped like a C or E, with gaps inside them that can fit other code blocks. These are called C blocks. The ●[If <condition> Then], ●[If <Condition> Then {} Else {}] are examples of C blocks. Code that is placed in the gaps of a C block is considered "nested" inside it in computer science terms. C blocks will exert some control on how the code placed inside them runs. Most C blocks are in the ●Control category.

Canvas – In Scratch, the *canvas* is part of the art workspace where you can actually draw and create images. The canvas has a centre point indicator and a line marking the 480 × 360 pixel area that will cover the entire Stage

Window, but it extends past this area. In general computing, the term *canvas* can apply to anywhere you can draw or show images, but its exact use may be more precise depending on the platform/context.

Cap Block – In Scratch, any code block that ends a script (it does not have a bottom connector, so no code can go below it) is a cap block. The most common cap block is ●**[Stop [All]]** in the ●Control category. The ●**[Forever {}]** code block is both a C block and a cap block because it repeats its nested code blocks endlessly and nothing can ever come after it.

Character – In computers, the term *character* can refer to either a typographic symbol, like a letter, number, or punctuation mark, or to a fictional being in a game, interactive novel, or other digital media. A character in a game could be a player character (PC), one that is controlled by a human player, or a non-player character (NPC) that is controlled by the computer. See also *player*.

Clone – In Scratch, a *clone* is a copy of a sprite. They function just like the sprite they are copying, but they never run the ●**[When ▷ Clicked]** event since they are created while the project is running and therefor do not exist when that event is triggered. They do, however, get access to the ●**[When I Start As A Clone]** event, which the original sprite they are duplicating will never be affected by. Clones are a very powerful technique to use in Scratch for populating a busy world or creating special effects, but Scratch is likely to crash if you create more than 300 clones at a time. Clones are dealt with through a trio of code blocks in the ●Control category.

Cloud – In general computing, *cloud* systems are massively scaled and redundancy-protected systems to handle user data or even run programming. The online version of Scratch is a cloud system, saving user accounts and data on massive web servers that users access through the Internet.

Code Block – A compartmentalized instruction for a computer system and fundamental unit of block coding. Each code block has a specific function that it makes the computer perform. Connected together in the right order, they can create algorithms and programs, giving full complex instructions for the computer to fulfil a given series of tasks. Each code block's shape and colour will provide users with details about how they connect and what its function might be, or how to find it in Scratch's editor. Keep in mind: code blocks are not unique to Scratch, and there are many coding websites and programs that have adopted using them as the means to program.

Code Block Library – In Scratch, when the Code tab is selected, the Code Block Library is displayed along the left-hand side of the editor. It lists all the code blocks available to the user to use in their project. They can scroll up or down through the list, or along the very left-hand side of the editor, they can click on any of the categories (•Motion, •Looks, •Sounds, •Events, •Control, •Sensing, •Operators, •Variables, or •My Blocks) of code blocks to jump to that category. The extensions add new categories and code blocks to the Code Block Library for that project.

Code Tab – In Scratch, this tab in the upper left-hand corner of the editor gives users access to the coding workspace in the middle of the editor as well as the Code Library along the left-hand side of the editor. In the Code tab, users can access the code blocks and create code for sprites and the stage. To make art, the Costumes tab or Backdrops tab are selected. To make sound, there's the Sounds tab.

Coding Workspace – In Scratch, when the Code tab is selected, the centre of the editor provides a workspace to do coding. Here, the user can arrange code blocks they drag from the Code Block Library to the left or by copying them in. The coding workspace is infinite, and users can click on blank areas of the workspace to drag it and change their view, creating more room as their projects grow. The code shown in the coding workspace is specific to the current selected sprite or stage. In the upper right-hand corner of the coding workspace, there will be a ghost image or watermark of the sprite or stage being coded for.

Collision Mask – In computing, a collision mask is a graphic that determines the positions that an object occupies and can collide or overlap with other objects. It may or may not match the size and shape of the object's display graphic. This method allows objects to have different visual looks compared to their collision behaviour, so in games objects can have artistic perspective that doesn't cause collisions despite literal graphic overlap.

Colour – In general, *colour* for digital systems is defined by three numbers assigned to red, green, and blue, or RGB. These control how bright each component light, affecting how a pixel on a digital screen looks, is turned on. If all three numbers are 0, the pixel is black, and if all three are maximum (generally 255) the pixel is white. Different combinations of values on each primary colour create different colours. In Scratch, *colour* refers to an exact output colour, what a pixel is coloured, but also refers to the hue of the final output colour, which is also affected by its saturation (how vivid or pale the output colour should be) and brightness (how bright or dark the output colour should be). It can also refer to an argument for the •Looks •[Change [Graphical Effect] By (#)] or •[Set [Graphical Effect] By (#)] code blocks that are used to shift the colouring of a sprite's costume, as opposed to the other possible graphical effects (like ghost or pixelate).

Colour Panel – The colour panel is a pop-up in Scratch that allows users to select a colour or fill pattern. It is used in numerous places wherever colour selecting is needed. In some places, some features of the colour panel are not available, depending on the context. Tips on working with the colour panel are in Chapter 4, "Scratch Basics".

Colour Picker – The colour picker is a tool accessed through the colour panel in Scratch represented by an eyedropper icon. It allows users to select a colour already in use, either in the canvas or the Stage Window, depending on the context. It will provide a zoomed-in view of an area and allow a single pixel to be clicked to provide that colour to the colour panel.

Combo – An unofficial term in block coding for whenever a code block is combined with another, especially when a Boolean or reporter code block, or code blocks, are placed into another code block. We also use the term *combo* to refer to some commonly connected code blocks, such as •**[When ▷ Clicked]** followed by a •**[Forever {}]**.

Comment – In general computing, a *comment* is some writing in a program that is not seen by the user or interpreted by the computer but rather simply acts to inform anyone reading the code of something. Commenting code is a good practice that can help make it more readable by providing titles, declarations, explanations, and footnotes to the programming. In Scratch, comments can be added to either the coding workspace (by right-clicking on it) or to a specific code block (by right-clicking on it). Then, a sticky note graphic will appear. It can be moved, scaled, and typed into to provide a comment.

Computational Thinking – In general education, the concept of computational thinking helps define a system for problem-solving. It is comprised of decomposition, abstraction, algorithm building, pattern finding, modelling and simulation, and evaluation. It provides practical methods to solve problems through both analysis, planning, and execution. See also *decomposition, abstraction, algorithm building, pattern finding, modelling and simulation,* and *evaluation*

Conditional – A *conditional* is a system of evaluation in computing. One or more metrics with one or more evaluations made of it returns a Boolean signalling if the condition is either true or false. For example, to check the weather in code could include •**[If** •<•**(Raining) = (True)> Then]**. Conditionals can also be numerical, as in •<•**(Temperature) > (32)>** or any other evaluation or comparison you might need on your project. In Scratch, conditionals are used with •Control category code blocks to make code run only if certain conditions are met so projects can be reactive and dynamic.

Costume – In Scratch, this is the graphic displayed to represent a sprite in the Stage Window.

Costume Library – In Scratch, if the user is in the Costumes tab, they can click the Add a Costume button to add a costume from the Costume Library. This lists all the graphics available by the default user.

Costume List – In Scratch, if the user is in the Costumes tab, all the costumes currently assigned to the selected sprite are shown on the left-hand side of the editor. The user can scroll up or down through the list. The selected costume displays in the art workspace. The user can click on any thumbnail in the Costume List to select it, and if desired, they can drag it up or down to change the order of the costumes in the Costume List.

Costumes Tab – In Scratch, the Costumes tab is available in the upper left-hand corner of the editor if the user has selected a sprite; otherwise, the stage selected and the Backdrops Tab will show in its place. Clicking on the Costumes tab allows the user to access the Costumes List and the art workspace to edit or create new costumes for the sprite they have selected.

Creator – In general, a *creator* is anyone using a program, or skill, to create something. In Scratch, it can mean the person using Scratch, the user of the editor, also known as a "Scratcher", or the person who created a project or asset that is being remixed. Because projects can be remixes of remixes or include assets from other projects or from outside Scratch, what exact relation a creator has to a project can be complex.

Decomposition – In education, *decomposition* is one of the components of computational thinking. It is based on breaking up a problem into distinct components to make smaller and easier tasks to deal with. This is useful in all problem-solving but particularly handy in coding because we can divide complex needs into discrete goals that can be easier to analyze and build solutions for, leading toward the end goal.

Define – One of the steps in the design thinking product design process. It comes after the empathise step and before the ideate step. It is where one takes the information and feedback from the empathy phase and identify a key problem and define it as an opportunity to brainstorm solutions to in the ideate phase. See *design thinking*.

Design Thinking – A design process model popular in the tech field. It is comprised of a cycle of five steps: empathize, define, ideate, prototype, evaluate/test. With product refinement achieved by rerunning the cycle. It reinforces the need of the user as the primary drive for design and ensures a strong link with the user through the design process. See also *empathize*, *define*, *ideate*, *prototype*, and *evaluating/testing*.

Direction – Sprites in Scratch have a direction property. This tracks or determines what angle they will move on if they **[Move (#) Steps]** and, depending on the rotation style property for the sprite, may orient their graphic to that degree angle as well. In Scratch, direction 0 is up, +90 is right, -90 is left. Sprites start with a default value of direction +90, or going right.

Duplicate – In Scratch, you can copy and paste code blocks, costumes, sprites, sounds, and backdrops. If you right-click on an object, you will get a pop-up of additional commands, generally allowing duplicating or deleting, but possibly others too. When you duplicate a code block, all code attached inside or under it will duplicate, too, and any typed or selected values will be copied as well. Duplicating a sprite will copy all its costumes, sounds, and code. There are many ways to copy things in Scratch. We give further explanation about it in Chapter 5, "Project 1: Dinosaur Dance Party".

Edge – In Scratch, the limit of the Stage Window view is called the edge. Objects collide with the edge the second any pixel of their costume touches it. The •**[If On Edge, Bounce]** code block will make an object that touches an edge bounce off it using the correct reflective angle. In many cases of basic movement, a sprite cannot move past the edge of the Stage Window. It will only move so far as some portion of its sprite is still visible in the Stage Window. This protection protocol can be worked around using Set X/Y movement, but •**[Move (#) Steps]** will be limited by it. The •**[Go To [*random position*]]** code block may move a sprite to colliding with the edge but will never place a sprite beyond the edge outside of the Stage Window. See also *Stage Window*.

Editor – The Scratch editor (or just editor for short) allows a user to create a new Scratch project or edit an existing one. It comprises of the Stage Window, stage, Properties panel, Sprite List, and then the left side of the screen depends on the tab selected, providing a workspace and tools for that tab. While the Scratch website also has the main page, search pages, studio pages, project pages, and others, we deal almost exclusively with the editor in this book because that's where all the code and digital art creation takes place. In general computing, an *editor* is any program that allows users to create or edit files, such as a word processing program, and is not limited to coding, but it can also apply to other software for coding.

Empathize – One of the steps in the design thinking product design process. *Empathize* is the phase where designers work with potential user groups to hear their experiences, problems, and needs. This gathers the information needed for the next step, *define*, where those problems are refined as opportunities for development. See also *design thinking*.

Evaluating/Testing – One of the steps in the design thinking product design process. *Evaluating/testing* is the final step, occurring after the prototype stage, where a product or service has been created and rolled out. Evaluating/testing determines how the product or service performs in the real world. Products can be further refined and developed by repeating the cycle, going to another empathizing phase to gather feedback from users after the evaluating/testing phase. See also *design thinking*.

Evaluation – In education, this is one of the six components of computational thinking. It is used to determine the effectiveness of a process or solution and to see if it can be generalized and applied to other problems. See also *computational thinking*.

Event – In Scratch, Events are a category of code block. Most events are hat blocks. Events are generally triggers that determine when code will be run. Without being attached to a trigger, code blocks will not be run. See *hat block* or *trigger*.

Execute – In computing, to *execute* means to run code. The computer is executing the commands you have given to it. This can also be referred to as run or running. In some programming languages, code must first be compiled (converted to direct instructions for the computer at the lowest level) before executing. Compiling is not necessary in Scratch.

Follow – Because online Scratch extends beyond just an editor, providing a platform for sharing content, user accounts can follow each other. When following another user, you get updates about their activity, such as when they share a new project. See also *Scratch account*.

Frame – In visual art (including video, animation, and video games), a *frame* is a single set complete drawing of the screen – all the data that comprises what is shown on the screen, or window, for a project. This is like a single picture in a film reel, or a single composited cell of animation. A frame represents the view of the project, animation, movie, or game world at that exact moment. A game, animation, or movie will have a frame rate that determines how rapidly a new frame is composed and drawn to the screen to create animation and update the user's view of the project.

Frame Rate – In visual arts, this is how quickly the graphical view of the project is updated, given in a number of frames per second (or FPS). Most computer imaging is done at 30 fps, or 30 frames drawn per second, though modern gaming has moved to 60 fps as a common standard. Cinema tends to run at 24 to 30 fps. The human eye and brain are believed to have a limit of comprehension and perception in the 30–60 fps range. This means we might not reliably notice things faster than that.

Function – In general computing, a *function* is a clearly defined set of instructions for the computer to achieve some particular purpose. It can refer to a built-in capability of a programming language, such as Scratch's ●[Go To [*Random Position*]], which automatically determines a randomized X-coordinate in a range of -240 to 240, a randomized Y-coordinate in a range of -180 to 180, and then changes the sprite's X and Y properties to those new numbers. Functions can also be user-defined or custom, where the user creating a program can define a new function and provide all the instructions for it. This function can then be called in their code whenever needed, just like a prebuilt function can. The ●My Block category in Scratch is the equivalent of building custom functions through the ●[Define [My Block]] hat block and calling them through the custom stack block created for any ●My Block that is created. Functions can have arguments or parameters that ensure needed numbers for the function to work are provided for it when it is called.

Grid – All computer graphical systems use graphs or grids to track the position and scale of objects they need to draw (show on screen). In Scratch, a coordinate system is used for the same purpose. It has its origin at the direct centre of the Stage Window, known as X:0 Y:0, and will centre a sprite. The Stage Window is 480 × 360 pixels, which gives visible coordinates from X: -240 Y: -180 to X: 240 Y: 180. X increases to the right, while Y increases to the top. Other platforms and digital systems may use different grid systems. One of the backdrops called "XY-grid" in the Backdrops Library can be used to see a diagram of the grid system in Scratch.

Hat Block – In Scratch, *hat blocks* are a shape of code block. They have a bump on the top that indicates nothing can attach above them. Hat blocks are all triggers, determining when the code attached below them should run. Most hat blocks are in the ●Events category. Every script must start with a hat block to tell the computer when that code should be run. Without a hat block, the computer will never run the code. See also *event* or *trigger*.

Ideate – This is one of the steps in the design thinking design process. It follows the define phase, where the challenges or problems have been identified and defined as opportunities. *Ideate* is where possible solutions are conceptualized to meet the opportunities. After ideate, the prototype phase builds systems imagined by the ideate phase. See also *design thinking*.

Input – In general computing, input is feedback from the user or environment. It can be direct commands, such as key presses or mouse clicks; indirect commands, such as mouse movement; or passive input, such as sensor readings or clock ticks. Input is used to interact with the computer system, generally triggering some code or providing values or data.

Lag – In general computing, this slang is used to refer to delays in processes. Most commonly, it refers to any pause, disruption to Internet streaming for games, videos, or video conferencing. It can also refer to delays in stimulus–response systems or delays between decisions and results. Scratch can lag if you have too many clones running overly complicated code.

Language – In general computing, this can refer to either a user's human language or a programming language, a system developed to allow humans to give instructions to the computer that is more readable and convenient than giving more direct and literal machine instructions (also known as assembly). In Scratch, *language* refers to user language. In the editor, you can change the language Scratch uses by clicking on the "Settings" button in the upper left-hand corner of the screen. Scratch is available in 74 languages at the time of writing this.

Message – In Scratch, a *message* is part of the broadcast system. Messages are uniquely named user-defined triggers that can be used through three of the Event category code blocks. Messages are broadcast – a sprite broadcasts the message using the **[Broadcast [Message]]** code block, and then any and all sprites or stage with **[When I Receive [Message]]** code blocks will trigger and run their attached code. This is a very useful system to get multiple objects to respond to a specific event.

Modelling and Simulation – In education, modelling and simulation is one of the six components of computational thinking. It's about using models and simulations as both a method of understanding situations and processes as well as creating or changing situations and processes. Coding is an excellent way to utilize this method, as we can not only create models and simulations but also can run easily and efficiently even at massive scales, from which we can derive results, but we can also easily and affordably run enough simulations to derive statistical data for further analysis. See also *computational thinking*.

Nested – In general computing, nesting code is a method involved in certain functions where some code is run conditionally or in some way controlled by another function. It is contained by this controlling function and, in typed code, is generally indented from its controlling agent, which nests it inside, making it visually distinct in addition to its flow control aspect. In mathematics, we use parenthesis to nest certain math functions inside others to control the order of operations, similar to how nesting functions in coding. In Scratch, the C blocks nest code inside them, in some way controlling how the code inside them is run, such as conditionally with an **[If <*True*> Then]** code block or loop it like with a **[Forever]** code block.

Object – In general computing, objects are used in some programming languages as a model for how code is run and data is organized. An object is a discrete agent in the model, with properties that can be set or checked and have code run by it or affecting it. Infinite numbers of objects can exist in a model, and they can have different properties and behaviours. Scratch uses this system, with both sprites and the stage being an object. They each have properties, can have their own code, and can be affected by or affect others.

Parameter – In general computing, a *parameter* is a data point that helps define how a function should be run or is another form of data point that defines how the program, model, or simulation works. In Scratch, the Turbo Mode setting would be a parameter, as would the View Modes. The term is sometimes applied to any argument, value, or input that is passed to a function.

Pattern Finding – In education, this is one of the six components of computational thinking. It is also sometimes labelled as pattern recognition. It is the practice of analyzing data or systems to find patterns, either in behaviour or outcomes, or in similar processes and outputs. Pattern finding is often used to help standardize or simplify processes or solutions. See also *computational thinking*.

Pen Tool – In Scratch, the Pen tool is one of the optional extensions that can be added to Scratch 3, creating a new Pen category of code blocks. It provides the ability to draw lines or copy costumes to the background, on top of the current backdrop. The pen works by being set **[Pen Down]** (to draw) or **[Pen Up]** (to not draw) and then moved. The code is run by a sprite so that a sprite's movement will determine what is drawn if they have the **[Pen Down]**. If a sprite uses the **[Stamp]** Pen category code block, it will copy their current costume to the background.

Pixel – In general computing, a pixel is the smallest unit of graphical display, equivalent to a single point of colour. Computer monitors, TVs, and other digital displays are comprised of millions of pixels that display images by lighting up different colours. Digital art can be made in two different methods: pixel (defined by a grid of pixel colours) or vector (defined by mathematical instructions to construct shapes with given properties). All art is always displayed on the pixels of a digital display, though. In Scratch, the **[Move (#) Steps]** block will move an object approximately (#) pixels on the screen, though this can be rounded off because of the angle of movement (see *direction*). The Stage Window in Scratch is 480 × 360 pixels, though in Presentation Mode, it is enlarged, so 1 pixel in-game will be multiple pixels on your display. See also *vector* and *grid*.

Player – In gaming, a *player* can mean either any top-level agent in the game capable of winning or scoring or, more specifically, a human playing the game. Players can be represented with avatars, or through characters, which are fictional agents within the game controlled by players. You can think of players as actors, and characters as the role the player takes in the game world. See also *character*.

Project – In Scratch, users create projects as the distinct applications or programs they can run and share. Each unique project is listed in the user's Scratch account. Each project gets its own project page, may or may not be shared, and can be edited in the editor and remixed or seen inside if shared. See also *project page* and *Scratch account*.

Project Page – Each project created in the online version of Scratch gets its own website page on the Scratch website. This page allows the creator to add instructions and notes to the project and to share it. If shared, the page can collect likes, favourites, and comments and allow others to play it, see inside the project, or remix it for their own use. A user's Scratch account links to all the project pages for all the projects they've created and saved.

Properties Panel – In the Scratch editor, the Properties panel is directly below the Stage Window and above the Sprite Listing. This panel allows you to see and edit the most commonly needed properties of a sprite. It has dialogs for showing/editing the name, X position, Y position, visibility, size, and direction of sprites, though sprites do have more properties than those listed.

Property – In computer science, a *property* is a piece of data assigned or associated with an object. It defines some aspect of its state in the simulation. Sprites in Scratch have many properties: name, X position, Y position, visibility, size, direction, costume, and more. Coding allows a creator to assess and alter properties of objects in their simulation, making them move, change colour, appear, or disappear. See also *object*.

Prototype – This is one of the five steps in the design thinking design process. It comes after the ideate phase, where the conceptual idea for a solution or product has been created. The prototype phase is the actual building of the product or service to bring the ideate phase's idea to life. It is followed by the evaluation/testing phase, where the prototype is tested in the real world to see how it performs. See also *design thinking*.

Remix – All projects in Scratch are under an open-source license as terms of use of the platform. This means that anything you create in Scratch can be shared with the world with no commercial use or restrictions on sharing. If you log in to a Scratch account, any project page you visit will have a Remix button on the top right corner. Remixing makes a copy of that project to your account that you can then edit and modify in any way you like, with the ability to share that project. This allows everyone on Scratch to learn from everyone else on Scratch since they can access the code they used to achieve their projects. We can all learn from everyone else through any shared project. See also *see inside*.

Reporter Block – In Scratch, reporter blocks are one of the shapes of code blocks. These are the round-edged pill-shaped blocks. They represent values, numbers, strings, or data. They allow a creator to refer to a data point somewhere in their project, such as ●**(X Position)** or ●**([direction] of [sprite1])**. They can also be typed-in data, such as ●**((#) x (#))**. Any of the white oval spaces in any other code block can be filled either through typing in a value or through using a reporter block.

Reticle – In the Art tab, the canvas has its centre point marked by a reticle, or target symbol. It can be important to align your costumes in reference to the reticle, as this is the point which a sprite is based. The X and Y properties will align directly with the reticle position of the costume. As well when rotating, the rotation of the sprite will be centred on the reticle. When moving vector shapes (or groups) on the canvas, they will automatically snap to the reticle if they are brought close enough to assist with properly aligning things.

Scratch – Scratch is a coding platform built with an emphasis on primary education and interactive media. It was developed by the Media Lab at MIT and is currently maintained by the Scratch Foundation. Scratch uses block coding to allow easy, friendly access for even young students. It is not just a platform for creation but also for sharing through the use of user accounts, studios, project pages, and a searchable listing of all shared projects created in Scratch. It is currently in its third iteration, Scratch 3. You can try Scratch at http://scratch.mit.edu/.

Scratch Account – The Scratch website is not just a platform for creation but also a platform for sharing. It has an account system for users to organize all their projects, save them on the cloud, and share them with the world. Scratch accounts are not required to use Scratch but are a great benefit to users as they provide a cloud saving for projects, an autosave feature (when logged in), and the opportunity to share projects with others.

Scratcher – A creator that uses Scratch can be referred to as a *Scratcher*.

Screen Refresh – In general computing terms, *screen refresh* refers to clearing the draw buffer in an operating system and rebuilding the graphical data to display to the user. Typically, this is done at every 30th of a second or more on a computer. In Scratch, "Run Without Screen Refresh" is an option available when creating a ●My Block. If checked, the code runs differently than usual and will not wait one frame between repeats and other similar delays in normal code. This can allow you to create custom functions with ●My Blocks that can rapidly process data, set up levels, or other things. See also *frame, frame rate*.

Script – In general computing, a *script* is a separate sequence of code that can be called as needed from any other code in a project and is a way to compartmentalize the reused functions required for a program. In Scratch, a *script* is any stack of connected code blocks. A script must start with a hat block so that the computer knows when to run that code due to a triggering event.

See Inside – In Scratch, every shared project can be viewed inside and out by other users. On any project page, there's a See Inside button in the upper right-hand corner. If clicked on, it will open the Scratch editor with a copy of the project loaded. Users can see all the code, sprites, costumes, backgrounds, and sounds that were used to make the project. This allows users to learn how to make any of the projects they explore, turning the site's ability to share into the ability to have millions of users teach one another.

Shared – Projects in Scratch can either be shared or not shared by their creator. A shared project can be seen by other users on Scratch, who can find it in search results, visit its project page, and even see inside the project or remix it. An unshared project cannot be seen by anyone except the user that created it.

Size – In Scratch, all sprites have a size property that determines how large they display in the Stage Window. Size is a percentile, with 100 indicating normal size relative to its current costume's size, 50 indicating half size, and 200 indicating double size. Costumes can be checked for their base scale in the Costume tab's art workspace.

Sound Library – In Scratch, when the user has selected the Sound tab in the editor, the Sound Library can be accessed by clicking on the "Choose a Sound" button in the bottom left-hand corner. The Sound Library is a collection of music, notes, and sound effects available to users to use built-in to Scratch. In the Sound Library, sounds are listed as tiles with a name and a play button that if the user hovers their mouse over it will play the sound effect. In addition, at the top of the Sound Library, there is a search bar and categories that will only list associated sounds. Clicking on a sound will add it to the currently selected sprite (or stage) in the project.

Sound Listing – In Scratch, when the user has selected the Sound tab in the editor, the Sound Listing is shown on the very left-hand side of the editor. It lists all the sounds currently added to the selected sprite (or stage). The user can select any of the sounds by clicking on them and can then edit or listen to them in the workspace. The user can click, drag, and reorder them in the Sound Listing.

Sounds Tab – In the Scratch editor, this is the third of the three tabs. It allows users to add sound effects and music to the currently selected sprite or stage or to edit already-added sounds. The workspace in the Sounds tab allows to play the sound clips as well as apply a number of different transforms or effects to either the whole sound or to portions of it by clicking on the sound wave to indicate starts and stops. The Sounds tab is explained in detail in Chapter 4 in Book 1 – Beginners.

Sprite – In Scratch, sprites are the main workhorse of projects. They are the individual objects that appear in the project through their costumes, take action through their associated code, and make sound with their associated sounds. Each sprite is its own discrete agent operating code-independently but can interact and react to the other sprites. All the sprites in a project are listed in the Sprite Listing, while the currently selected sprite displays its properties in the Properties panel. Only one sprite, or the stage, can be selected at a time, and whatever is selected is the target for any work done in the workspace, whether code, art, or sound. In general computing, the term *sprite* refers to a piece of pixel art that will be displayed to represent an object, rather than being an object in its own right with swappable costumes.

Sprite Library – In Scratch, if the user decides to add a new sprite, they can click on the Choose a Sprite button in the bottom right-hand corner of the editor. This will show the Sprite Library, a collection of already-drawn objects available to all users of Scratch. In the Sprite Library, if a user hovers their mouse over one of the tiles, they may see it animate, showing the multiple costumes associated with that chosen sprite, but not all sprites have multiple costumes, so not all will animate.

Sprite Listing – In Scratch, all the sprites that have been added to a project appear in the Sprite Listing. This is found under the Properties panel, which is under the Stage Window in the bottom right-hand corner of the editor. Each sprite in the game is listed here in tiles, with a thumbnail displaying the current costume active for that sprite. The user can right-click on a tile to duplicate it or delete it. They can also click and drag sprites to reorder them in the Sprite Listing. The currently selected sprite will be outlined in blue, and a trash bin icon will appear in the upper right-hand corner of the tile that can be clicked to delete it. Whatever sprite is currently selected will display its properties in the Properties panel above, as well as show its associated code, costumes, or sounds in the workspace to the left.

Stack – In Scratch, a *stack* is a set of interconnected code blocks. The official name for a stack is "script", but *stack* is very commonly used. See also *script*.

Stack Block – In Scratch, the stack block is the most common shape of code block. These are the basic rectangular code blocks. They can fit above or below other code blocks and often have a value space that the user can type in a number or text into or can use a reporter block to fill to have it populate with a dynamic value from the project. Stack blocks connect into scripts, and each script must start with a hat block in order for the computer to know when to run the code.

Stage – In Scratch, the *stage* is a special object that handles the background in projects. It can have its own code, graphics (called backdrops), and sounds. The stage has limited code blocks that can be assigned to it because it cannot move, change layering or size, among other limitations. It is always directly centred in the Stage Window and is always the exact same size. The stage cannot be removed or deleted from a project.

Stage Window – In Scratch, a Project is displayed to users through the Stage Window. This appears in both the project page and in the editor. The Stage Window is a view into the simulation created in Scratch, whatever form it has taken – game, music video, interactive story, etc. The stage will always provide the background of the Stage Window since neither the view the Stage Window provides nor the stage is able to move. The Stage Window is 480 pixels wide and 360 pixels tall. Its border is called the edge. Whenever the project is played (through clicking the ▷), all the action and interaction will occur in the Stage Window. See also *edge*, *grid*, *project*.

STEAM – In education, STEAM is an acronym standing for science, technology, engineering, art, and math. It represents the earlier term STEM with the addition of *arts* to highlight the need for creativity. STEAM and STEM are terms associate with a push to highlight knowledge of the physical sciences and applied science in both formal education and after-school programs or hobbies. Coding is often used as a way to incorporate technology and/or engineering into STEAM curriculum.

STEM – In education, STEM is an acronym for science, technology, engineering, and mathematics. It is a common term used to highlight a focus on physical and applied science in education. Coding education has often been used as a major focus for STEM programs as a way to incorporate both technology and engineering. STEM is sometimes extended to STEAM with the addition of *art* as an additional focus.

Step – In Scratch, a *step* is a measure of distance, roughly equivalent to 1 pixel. In computing, it will often be used as a reference to the execution of the sequential order of code, with each line of code (or code block in Scratch) equating one step of execution.

String – In general computing, a *string* refers to a sequence of characters or typographic symbols. Words, sentences, and passwords are all strings. They are a form of variable or data point that is not numeric – therefore cannot have math operations performed on it. See also *value* or *variable*.

Touching – In Scratch, sprites can be tested if they are touching other sprites or colours, or if colours are touching other colours. *Touching* in this case is determined by the costumes of sprites involved, or the sprite involved and the colour of the background, or any other sprites as composited into a frame at that time. It is determined if any pixel, or pixels, of the assigned colours overlap in position. In general computing, *touching* is known as colliding. Importantly, touching only counts overlapping, not adjacency.

Trigger – In general terms, a *trigger* is anything that sets into motion an action or reaction. In coding, this is often an input (such as a key press) from the user or an input from a sensor system. The hat blocks in Scratch are examples of triggers; they are specific events that will cause Scratch to recognize their occurrence and can then be used to have the computer run associated code assigned under them. See also *hat block*, *event*, or *input*.

User – In general terms, a user is anyone using an application, program, or project. This can refer to either the person playing a project in Scratch (who can also be referred to as a player) or as the person using Scratch to create a project (who can also be referred to as a creator or Scratcher). See also *creator* or *player*.

Values – In general computing, a *value* is a data point. It can be assigned to a variable or required for a function or found in a data structure. The term is used widely and freely to refer to any kind of data point and may refer to an input or parameter. In Scratch, a value is a white oval in a code block, a data point required for the code block to do its job, helping define or quantify its actions. Values can be represented through reporter blocks, and reporter blocks can fit into value places in code blocks. See also *input*, *parameter*, *variable*, or *reporter block*.

Variable – In general computing, a *variable* is a memory assignment for the computer. It creates a reference name to a piece of data that the computer will hold in memory, which can be changed or referred to at will. By making a general reference, this data can be referred to at any point or can be modified as needed. In some computer languages, variables have set data types – integer or string, for example; in others, they are dynamic. In Scratch, ●Variables are used the same as in general programming as fully dynamic data points, but it is also a category of code blocks that are used to work with ●Variables. To work with a variable in Scratch, one must first go to the ●Variables category and click on the Make a Variable button to create it first. See also *string*.

Vector – In general computing, *vector* art is one of two forms of art, the other being pixel (also known as raster). In pixel art, the canvas is defined as a grid of cells, with each cell representing a single pixel with an assigned colour. In vector art, the art is defined by mathematical formulae and instructions on how to build dynamic relationally positioned and proportioned shapes, lines, and spaces. Vector art, therefore, scales perfectly without any loss of quality and can have true curves and smoothness, unlike pixel art. It is, however, more difficult to create complex designs with. Scratch can create and edit both forms of art. It is highly recommended you try learning both and encourage students to do the same. We talk more about the two forms of digital art in Chapter 4 in Book 1 – Beginners. See also *pixel*.

View Mode – Scratch has multiple ways of presenting projects to users. By default, either on the editor or the project page, you will see the Stage Window at normal scale. In the editor, you can also choose between three view modes – default, compact (the Stage Window is half size, so you have more room to code), or presentation, also known as full screen. In Full Screen mode, the Stage Window will expand to fill as much room on your screen as possible. You can access the Default or Full Screen mode in either the editor or the project page. Full Screen mode can help students remember that they can't count on users being able to click on code blocks to execute them and must make full controls that don't rely on the user being able to manipulate code or sprites as an editor. See also *Stage Window*.

Workspace – In the Scratch editor, the workspace is the largest central part of the editor. Depending on which tab is selected, it is where the user can create their code, create or edit art for costumes or backdrops, or listen or edit sound. It will change nature with appropriate tools, depending on the tab selected. The content of the workspace will depend on the sprite or stage selected. See also *art workspace* or *coding workspace*.

X – In Scratch, X is both a position property of sprites and a dimension of the grid used to determine position and scale within the Stage Window. X measures the left–right positioning. X: 0 is the centre of the Stage Window, with X increasing to the right. The Stage Window is 480 pixels wide, making grid positions range from X: –240 to X: +240. See also *Y*, *grid*, or *Stage Window*.

Y – In Scratch, Y is both a position property of sprites and a dimension of the grid used to determine position and scale within the Stage Window. Y measures the up–down positioning. Y: 0 is the centre of the Stage Window, with Y increasing to the top. The Stage Window is 360 pixels tall, making grid positions range from Y: –180 to Y: +180. See also *X*, *grid*, or *Stage Window*.

Index

Green Text – entry is in Book 1
 – Beginners
Blue Text – entry is in Book 2
 – Intermediate
Red Text – entry is in Book 3
 – Advanced

A

accessibility 16–17, 19–21, 7–8
accounts 21, 31, 153–154, 158, 193–195
add a background 69–70, 87–89, 24, 39–40
add a costume 90, 114, 58, 64
add a sound 61, 65, 71–72, 77, 80, 98, 131–133, 113–114, 64–65, 138–140, 146–150
add a sprite 47–48, 72, 89, 104, 106, 112, 23–24, 26–27, 45–48
ADD/ADHD *see* sensory issues
advanced students 148, 155, 151, 159–160, 184, 191, 196
AI 1, 11, 24, 76–78, 83–85, 108–109, 131–134, 142–143
algorithms: flow of code 146–147; grid marching 26–29; list stepping 17–19; place sorting 154–158
ammo/use limits 123–124, 138, 165–167
analogies, helpful 110–111, 60, 149–150
and logic 115–116, 52–55, 61–62
animation: cycle 72–74, 82–83, 34–35, 110–111, 128–131, 163–165; frame 62, 72–74, 78, 104–105, 121, 42, 52–53, 58–59; Fx 99–100, 58–61, 119–123, 146–150; motion 78–79, 92–94, 105–106, 108, 48–49, 58–61, 65–68, 138–140; reactions 117–121, 75–79; techniques 95–97, 41–42, 45–48, 34–35, 138–140, 163–165

animation projects 68, 85, 11, 32, 28–68
answer *see* ask
art: colour effects 29–32, 82–84; costumes (*see* costumes); digital skills 22–24, 146–148, 163–165; formats 50–51, 62; glow effect 112–124; hidden objects 58–61; invisible components 17–19, 75–79, 125–127; making in scratch 47–56, 87–91, 106–108, 112–113, 114, 23–24, 26–27, 74–75, 80, 88–89, 98, 100, 104, 106–107, 112–113, 117–119, 122, 124, 128, 129–130, 16–17, 23–26, 36–38, 58–61, 65–68, 71–72, 75–107, 110–117, 119–123, 125–140, 146–161; making patterns 20–21, 21–23; rotating 49; shape 87, 107, 11; symmetry 107, 23–26, 112–114, 119–123; vector 88–91, 107, 23–24, 88–89, 98, 104, 65–68, 69–107; working with backgrounds 87–89, 24, 125–127; working with sprites 47–56, 89–91, 107–108, 112–113, 24, 26–27, 56–57, 58–60, 62–63, 87–88, 98–100, 106–107, 118–119, 121–122, 16–17, 23–26, 30–31, 36–38, 58–61, 65–68, 71–72, 75–107, 110–117, 119–123, 125–140, 146–161; working with text 112–113, 80, 100, 121, 65–68, 104–105, 105–107, 146–150, 151–153, 154–158, 158–161
Artificial Intelligence *see* AI
art projects 85, 11, 69–107, 108–161
art tools 51–54; *see also specific tool names*
ask 35–39, 43–45
assessment 149–150, 154–155, 150, 152, 159, 180–182, 184, 196
assignments 31–33, 144–145, 148, 151, 184

attacks 112–114, 128–131, 135–137, 140–142
autism *see* sensory issues
automation 11, 20–23, 12–13, 32–33, 176–179
autosave 154, 158, 193–195

B

back *see* layers
backdrop: tab 47, 59–61, 87–89, 24, 58–61; blackout 24, 110; setting 69–70, 87, 24, 39–40, 96–97, 30–31, 36–38, 38–44, 48–56, 58–68
background music 71–72, 83–84
backgrounds, scrolling 125–127, 163–165
backpack 76–77, 157, 158, 162, 193–195
back up plans 164–166, 169–170, 205
backward *see* layers
behaviour, code 47, 61, 155–157, 148–150, 159–162, 196–198
best practices: commenting 149–150, 152, 185–186; development aids 36–38; encapsulating 116, 60; initializing 105–106, 111, 8–9, 15–16, 34, 39, 91–94, 113–115, 123–124, 36–38, 79–81, 85–88, 110–101; instruction 155, 160–161, 196–198; reset 94–95, 14–15, 14–16; troubleshooting 152, 155, 157, 193–207; visibility 97–98, 117, 124–132, 114–117
bias 12, 16, 6–7, 7–8
biology 102, 117, 21–23, 176–179
bitmap 50–51
block coding concept 17–20
bonuses/boosts 96–98, 123–124, 93–94
Boolean 43–44, 46; *see also* if
boss enemies 138–145
bounce, if on edge 120, 114–117
brightness 54, 89
broadcasts *see* messages
brush tool/size 51–52, 90–91
bugs/bug hunting 27–28, 97–98, 152, 157, 182–183, 193–207
buttons 55–57, 124–127, 129–132, 142–143, 65–68, 151–153, 158–161

C

canvas, art 48–52, 54–55, 87–91, 107–109, 24, 104–105, 17–19, 75–79, 125–127
careers 10–11, 6–7, 142, 7–13, 108–109, 162, 180–181, 208–210
categories, code block 41–42, 41–47
center *see* reticle
challenges, for advanced students 148, 151, 9–11, 184, 191
change X/Y *see* X/Y
character 104–105, 32, 78–80, 98, 28–68, 30–31, 69–107, 72–73
choice, player/user 15–16, 43–45, 73–74, 85–86, 129–132, 28–68, 158–161
circle tool 51, 53, 74–75, 88–90, 119–123
clamping 85–86, 135–136
clipping, sounds 64–65
clones: child/parent difference 125, 164–165, 76, 168, 135–137, 204–205; clearing out 91–94, 96–98, 143–144, 112–117, 128–131; for difficulty scaling 123–124, 140–142; procedural generation 106–109, 83–85, 97–101, 114–117, 123–125, 128–131, 131–134, 140–142; for projectiles 71–72, 112–114, 128–131, 135–137; techniques 96–98, 143–144, 135–137, 175–176; troubleshooting 164–165, 168, 204–205
cloud saves 31, 154–158
cloud variables 108–109, 154–158, 165–167
code blocks 38–39, 41–47, 107–109, 146, 148, 14–16
code disappeared 40–41
code flow *see* flow of code
code for the stage 71–72, 39–40, 48–50
code tab 38, 39–40, 60
coding is for everyone 1–2, 10–14, 169–170, 6–7, 148–150, 173–174, 7–8, 179, 208–210

coding problems 154, 159–160, 196
coin toss *see* randomization
collaboration 149–150, 154–155, 150, 152, 159–160, 180–186, 193–207, 208–210
collectables *see* power-ups
collision-based movement 109–111, 94–97, 167–172
collision correction 117–121, 73–75, 94–97
collision masks 75–79, 169–172
collisions: among clones 96–8; with changing objects 97–103; colours for terrain 111–112; for gravity 73–74; invisible or hidden 75–79; mutual destruction 117–119; projectiles 76–78, 117–119, 128–131, 135–137; for transitions 38–44
colour 19, 21, 41–42, 46–47, 54, 89, 91, 163, 29–32, 111–112, 112–113, 135–137, 201–204, 82–84, 201–204
colour, pen extension 29–31
colour collisions 163, 111–113, 117–121, 139–140, 168, 204–205
colour effects 63, 29–31, 82–84
colour panel/picker 47, 54
colour selection 163, 111–113, 168, 204–205
commenting 27–28, 149–150, 152–153, 9–11, 185–186
community 2, 4, 25–28
comparison, numeric (<=>) 126–127, 131, 51–52, 71–72, 85–88, 115–116, 123–124, 32–35, 48–50, 51–62, 75–88, 94–101, 104–105, 112–161
comparison, text 35–38, 43–45, 38–44, 45–50
compositing, layers 54–55, 56, 62, 88–89, 98, 100, 104, 106–107, 121, 112–117, 119–123, 138–140, 146–150; *see also* grouping
computational thinking 14, 22
computer opponents *see* AI
computer science 22, 24–25, 6–7, 140, 146, 148–150, 173–174, 1–16, 19–21, 165–167, 176–179, 182–183, 208–210

concurrence 86, 140, 162, 91–94, 201–204, 201–204
conditional statements *see* if; until
confused pairs 157–160, 162–165, 198–201
constant learning 5–6, 13–15, 169–170, 6–10, 146, 173–174, 3, 176–179, 179–180, 208–210
control blocks *see* if; repeat; until
controls: dragging 136, 15–16; keyboard 77, 80, 83, 92, 95, 105, 13–20, 23, 25, 27–30, 35–36, 85–86, 106–111, 123–124, 138–139, 14–17, 36–38, 72–73, 74–75, 94–97, 110–111, 112–114; mouse 113, 121–122, 124, 52, 54–57, 71–73, 76, 100–101, 125–126, 129–132, 30–31, 36–44, 50–56, 65–68, 105–107, 151–161; text input 35–39, 139; variables 18–19, 21–23, 72–75, 129–132
control structures, troubleshooting 161, 166–167, 201–204
conversations 129–131, 45–48, 52–55, 64–66, 50–58
cooldown 121–124, 112–114
coordinates *see* X/Y
copying code 75–77, 93, 157, 161–162, 196–198
costumes: about 48, 58, 61–63, 65; altering 62; animating 72–74; collision masks 169–172; creating 89–91, 106–108, 23–24, 26–27, 74–75, 85–88, 97–101, 98–100, 100–101, 104–107, 112–113, 116–119, 121, 124, 128–130, 16–17, 23–26, 36–38, 58–61, 65–68, 71–72, 75–107, 110–117, 119–123, 125–140, 146–161; mixing sprites 56, 58, 63–64, 30–31; movement-based 75–79; randomizing 27–29, 89; size limit workaround 104–106, 17–19; as sprite property 139, 143; as states 95, 117–121, 69–70, 163–165; tab 47–56, 89–91, 104–105, 23–24, 146–148

creativity 4–6, 12–13, 22–23, 25–26, 62, 169–170, 146–148, 173–174
critical thinking 13–15, 22, 169–170, 4, 148–150, 173–174, 9–11, 180–183, 208–210
CS *see* computer science
CSV files 12–13, 27
Ctrl + C/X/Z 56, 63, 75
curved (vector) 48, 53, 87–89, 74–75

D

damage systems 115–121, 117–119, 128–131, 131–134, 140–142, 144–145
data: about 45–47, 57, 110–111, 137, 12–16; processing 171–172, 12–27, 165–167; visualization 12–27
deaths 109–110, 69–70, 82–88, 119–123, 163–165
defined count 18–23, 27
delete 39, 45, 52, 57, 63, 72, 86, 88
depth *see* compositing, layers
design thinking 13–14, 175–176, 180–181
dialogue *see* conversations
difficulty scaling *see* game balance
digital citizenship/literacy 5, 11–12, 6–7, 208–210
direction 37, 63–64, 120, 137–139, 14–17, 19, 48–50, 71–75, 83–85, 106–111, 63–64, 75–79, 97–101, 114–117, 128–140
displays 110–116, 18–19, 21–22, 56, 74–75, 80–83, 121–122, 124–132, 142–143, 12–27, 36–38, 85–88, 151–161, 163–165
download 27–28, 29, 31, 154, 169, 206
drag mode *see* controls, dragging
duplicate code 75, 93, 16, 21, 56–57, 61–62, 125–127
duplicating sprites 100, 56, 83–85, 143–144, 23–26, 38–44, 51–61, 91–92, 131–134, 154–161

E

edge 37–38, 107–110, 120, 125, 17–18, 109–111, 73–74, 82–84, 112–117, 125–127, 131–137

encapsulating *see* best practices, encapsulating
end game 110–111, 114–116, 125–131, 48–50, 66–68, 98–100, 116–117, 61–65, 10405, 146–150
enemies *see* opponents
engagement 14–15, 21, 154–155, 142–143, 146–150, 159–160, 7–8, 175–176, 196
environmental effects 105–106, 94–96, 111–112, 73–74
erase 51, 53, 14–15, 29–31, 119–123
errors: as assessment tool 154–155, 160–162, 164, 8–10, 157, 159–160, 165–169, 179–182, 196, 201–205; growth mindset 154–155, 164–166, 8–10, 135–136, 146–150, 157, 159–160, 165–169, 173–174, 179, 181–182
evaluation, 13–14, 182–183; *see also* errors, as assessment tool; if; until
events blocks 42, 47, 71
events, broadcast messages *see* messages
events, concepts 139–141, 146–147, 134–135, 140–141
exceptions 114–116, 125–131, 161, 8–10, 17–18, 35–38, 43–45, 51–52, 81–83, 85–86, 96–98, 104–106, 109–112, 115–121, 123–124, 135–136, 138–141, 143–144, 148–150, 157, 9–11, 21–22, 32–35, 45–48, 50–62, 73–79, 94–97, 112–114, 119–123, 181–182, 193–207
exploration 30, 144–145, 148, 154–155, 8–10, 151, 159–160, 173–174, 28–68, 69–107, 176–179, 196
explosions 95–96, 128–131, 146–150
extensions 24, 148, 12–13, 151

F

fade transitions 57, 99–100, 41–42, 48–50, 58–61, 80–81, 64–65, 85–88
false 44; *see also* if; until
fill 51–52, 54, 87–89, 89–91, 106–108, 112–113, 23–26, 74–75, 88–89, 98, 104, 17–19, 75–79, 114–117, 119–123, 125–127, 128–131

flow of code 47, 139–141, 146–147, 160–161, 148–150, 201–204, 180–182, 201–204
following 26–28
font 53, 112–113, 78
forever 71, 72, 106, 108, 119–120, 125, 41–42, 53, 74–75, 83–86, 106–109, 36–44, 65–68, 72–79, 85–92, 97–103, 110–123, 125–127, 131–142
forever, pausing/stopping 114–116, 125–129, 52–55, 81–83, 115–116, 82–84
forward (layering) *see* layers
frame animation 48, 72–74, 92–93, 139, 140, 160–161, 117–121, 165–167, 34–35, 79–81, 85–88, 94–97, 125–131, 163–165, 201–204
frame rate 74, 92–93, 141–142, 113–115, 117–121, 123–124, 138, 73–74, 94–97, 112–114, 119–123
freezing 153, 158, 193–195
front (layering) *see* layers
functions 42, 60–61, 81–83, 91–94, 135, 7–11, 94–97, 167–169

G

game balance 118, 83–85, 87–91, 96–98, 104, 111–112, 117–121, 143, 153, 155, 71–72, 82–88, 101–103, 108–109, 112–114, 138–140, 158–161, 165–167, 175–176, 187–189
game over 109–110, 114–116, 125–130, 8–10, 48–50, 66–68, 98–100, 116–117, 138, 140–141, 2, 82–84, 114–117, 119–127, 140–142, 144–150, 187
game projects 102, 117, 144–145, 32, 69, 102, 28–68, 69–107, 108–161
game start 71, 105–106, 108, 111, 113–115, 119–120, 122–126, 34, 38–39, 41–43, 91–101, 105–108, 112–114, 121–127, 48–50, 63–68, 79–81, 105–107, 110–123, 125–127, 138–140, 151–153, 158–161
game states: concept 125–131, 161, 80–85, 115–124, 140–141, 165–167, 171–172, 4–6, 36–38, 50–62, 82–92, 119–123, 146–161, 180–181, 201–204; damage 117–121, 82–84; lives/death/game over 109–110, 125–129, 130–131, 48–50, 66–68, 98–100, 116–117, 61–62, 82–88, 91–92, 104–105, 119–123, 146–150, 172–175; narrative paths 35–36, 43–48, 51–52, 55–57, 28–29, 36–38, 51–62; play & pause/turns 80–85, 115–122, 140–141, 69–70, 82–88; waypoints & levels 89–92, 104–105
gate systems *see* key/lock systems
generating enemies 123–124, 76–78, 83–85, 87–88, 143–144, 158, 168, 108–109, 114–117, 123–125, 128–134, 138–140, 165–167, 175–176, 193–195, 204–205
generator objects 123–124, 71–72, 76, 83–85, 96–98, 143–144, 97–101, 114–117, 123–125, 128–142, 165–167, 175–176
ghost effect 63, 99–100, 41–43, 48–50, 58–61, 80–81, 64–65, 85–88, 101–103
Glide 45, 67, 158, 162, 53–54, 162–163, 167, 30–33, 55–58, 63–64, 85–88, 138–143, 146–150, 198–201, 201–204
Go To 105–106, 108, 122–124, 20–23, 25, 27–28, 55–57, 74–75, 78–80, 83–94, 96–98, 105–107, 19–23, 38–48, 56–64, 75–84, 89–90, 94–97, 97–101, 110–117, 119–123, 125–153
Go To Vs Glide To 158–159, 162–163, 198–201
Go To X Layer *see* compositing
gradient 52, 54, 87–91, 106–107, 112, 114, 74–75, 88–89, 98, 100, 75–79, 112–117, 119–123, 128–131, 135–137
graphical effects 41, 99–100, 139, 143, 163, 4–5, 29–30, 42, 48–50, 58–61, 66–68, 80–81, 100–101, 143, 167–168, 2, 82–88, 101–103, 119–123, 125–127, 144–145, 163–165, 198–201, 204–205
graphics *see* art
gravity 105–106, 71–72, 4–6, 73–75, 167–172
green flag 37, 47, 71–72, 74, 112–113, 119–120, 123–124, 141, 15, 34, 45–47,

72–76, 85–94, 96–101, 106–107, 117, 121–122, 124–128, 168, 16–17, 23–26, 30–31, 36–44, 50, 52–55, 65–68, 71–72, 75–92, 94–97, 101–103, 105–107, 114–117, 128–131, 151–158, 204–205

green flag, restarting without 105–106, 112–113, 105–107, 112–117, 124–132, 110–111, 114–127, 146–153

grid *see* algorithms; grid marching; X/Y

grouping 54–55, 62

H

handles (vector) 48, 53, 104

hat blocks *see* events; messages

hazards 106–109, 123–128, 83–85, 98–100, 117–121, 82–88, 128–131, 135–137

health 109–110, 125–129, 115–121, 82–88, 114–123, 172–175

hide *see* visibility

high scores 110–111, 122–123, 131, 87–88, 96–98, 113–117, 117–119, 128–134, 144–145, 154–158

hit points *see* health

I

If On Edge, Bounce 119–120, 125, 114–117

If Statements: game state 125–131, 81–85, 115–116, 50–62, 79–105, 119–125, 138–140; keyboard controls (*see* controls; keyboard); mouse controls (*see* controls; mouse); multiple choice 43–45, 32–33, 38–44, 48–52, 55–62, 154–158; multiple conditions 43–45, 115–116, 52–55, 61–62, 73–74, 117–119; properties 71–72, 32–35, 38–44, 48–50, 52–55, 82–88, 112–119, 125–131, 144–145, 151–153; random chance 26–29, 135–137; sprite collisions 109–110, 76–78, 81–94, 98–100, 107–111, 38–48, 73–75, 82–88, 91–101, 117–123, 140–142; text comparison 35–38, 45–50, 52–55, 82–84, 97–101; variable assessment 125–131, 81–86,

115–124, 129–132, 21–22, 32–35, 45–48, 50–55, 58–62, 75–90, 94–101, 104–105, 112–114, 119–125, 138–150, 154–161

If Vs If/Else 43, 146–147, 158, 160, 8–10, 43–45, 165, 198–201

inertia & friction 73–74, 94–96, 111–112, 69–70, 72–73, 75–79, 167–169

inheritance *see* parent/child concept

initialize *see* game start

input *see* controls; keyboard & controls; mouse

interface *see* controls; displays; variable displays

internet problems 20–21, 153, 164–165, 158, 169–170, 7–8, 193–195, 205–206

inventory 45–57, 28–29, 36–38, 52–61, 163–167, 172–175

invisibility *see* visibility

Item Use 45–48, 52–66, 123–124, 36–38, 52–55, 64–65, 165–167, 172–175

iteration 140–141, 146–148, 8–10, 20–23, 135, 138, 94–97, 167–169, 175–176, 184

J

jobs *see* careers

join text 129–130, 38–39, 41–43

jumping 105–106, 69–70, 74–75, 93–94, 167–169

K

keyboard input *see* controls, keyboard; key press event

key/lock systems 45–48, 52–57, 140–141, 28–29, 36–38, 50–62, 172–175

key press, testing 85–86, 106–111, 138–139, 36–38, 72–75, 91–92, 94–97, 110–114

key press event 47, 77, 82–83, 94–95, 105, 141, 143–144, 14–17, 19–23, 25, 27–30, 123–124, 138–140, 14–19, 36–38

L

language 17, 18, 22, 26, 34, 149, 36–38, 143, 148–150, 152, 171–172, 1–3, 7–8,

12–16, 22–23, 182–183, 185–186, 208–210
layers 41, 48, 54–55, 63, 113, 163, 41–42, 48–50, 56, 60–61, 76–77, 98–101, 104–109, 121–122, 124–129, 142, 167–168, 30–31, 52–55, 65–68, 71–72, 75–81, 85–88, 91–92, 104–107, 114–117, 119–123, 125–127, 140–142, 144–145, 151–153, 163–165, 169–172, 181–182, 204–205; *see also* compositing
level-based clones 96–98, 143–144, 97–101, 172–176
level-based costumes 104–106, 129–132, 134–135, 140–141, 71–72, 82–84, 91–94, 101–103, 172–175
level-based properties 79–81, 89–92, 94–97, 172–175
level systems 104–106, 116–117, 129–132, 134–135, 140–144, 171–172, 4–6, 69–72, 79–81, 91–107, 162, 172–176, 181–182
limitations: ammo/power up uses 123–124, 165–167; applying variable 85–86, 48–50; canvas and art size 35, 37–38, 89–91, 106–109, 146, 17–18; clone count 153, 158, 193–195; positions 17–18, 48–50; size 104–106, 135–136, 17–19; touching options 96–8; visibility & collisions 75–9; *vs.* text code 7–8
Line tool 52, 23–26
list variables 12–22, 27, 165–167, 172–175
lives *see* health; variables, health/lives
logic *see* computer science; game states; operators
logic, teaching 13–14, 17, 19, 22, 42, 117, 147, 161–162, 169–170, 8–10, 140, 142, 148–150, 165–166, 173–174, 9–11, 154–158, 181–183, 201–204, 208–210
Login 21, 31, 153–154, 158, 193–195
lookalike code blocks 157–160, 162–165, 198–201

looks blocks, troubleshooting 159, 163, 164, 167–168, 198–201, 204–205
loops 42–43, 71–72, 78–79, 108–109, 130–131, 140, 36, 71–72, 112–113, 134–135, 32–33, 94–97; *see also* forever; music; repeat; until
loss *see* game over

M

Make a Variable 110–111, 122–123, 125–126, 18–19, 21–23, 14–16
masks 69–70, 75–79, 169–172
mathematics: about 22, 28, 42, 44, 46, 50, 110–111, 19, 133, 136–138, 19–21, 35–36, 175–176, 176–179; formulae 18–19, 21–26, 72–75, 83–91, 94–96, 106–111, 117–121, 17–26, 35–36, 38–44, 55–58, 73–79, 94–101, 110–111, 140–142; geometry 11–31, 71–75, 17–23, 35–36, 74–75, 97–101, 110–111, 114–117, 142–143; multiplication 79, 82–83, 92–93, 99–100, 105, 108, 21–29, 19–27, 74–75, 93–94
maximums 18–19, 85–86, 21–22, 48–50
menus 112–116, 8–10, 124–132, 134–138, 140–143, 150, 9–11, 65–68, 105–107, 151–153, 154–158, 158–161, 181–182
messages: about 46, 47, 149, 8–10, 35–36, 13–15, 140–141, 150, 152, 185–186; cutscenes 45–48, 52–55, 58–66, 117–123, 51–55, 89–90, 140–142; game over (*see* game over); generation 119–123, 128–131, 138–142; interfaces 58–66, 76–77, 81–83, 124–132, 142–143, 21–26, 151–161; sequential questions 35–36, 43–45; start game (*see* game start); transitions (*see* transitions); troubleshooting 159, 163, 159–160, 165–169, 196–198, 201–204
minimums 18–19, 21–23, 85–88, 135–138, 17–19, 21–22
missed shots, deleting 71–72, 76–78, 87–88, 112–114, 119–123, 128–131, 135–137

motion blocks: arc motion 71–75, 74–75, 167–169; centric movement 119–120, 125, 11–26, 71–72, 106–111, 117–121, 85–88, 167–169; dynamic speed 73–74, 94–96, 109–112, 117–121, 123–124; gravity 105–106, 71–72, 73–74; random movement 95–97, 120–125, 23–26, 36–37, 87–91, 96–98, 97–101; rotation styles 34, 104–109, 30–31; stage 48–49, 146, 155–156, 160–161, 82–84; using variable input 18–23, 72–73, 83–85, 17–21, 30–33, 38–44, 55–58, 72–75, 89–92, 167–169; X/Y movement 79–82, 92–94, 26–29, 78–80, 87–91, 17–21, 30–33, 55–58 72–75, 82–84, 89–92, 97–101, 167–169
mouse controls 113, 121–124, 126, 144, 15–16, 52–66, 71–72, 76–77, 100–101, 124–129, 30–31, 38–44, 169–172
mouse over 47, 70–71, 90, 106, 112, 136–137, 23–29, 45–48
movement: cycle lockout 109–110, 114–116, 126–127, 130–131, 76–78, 83–85, 87–88, 45–48, 63–64, 73–74, 85–88, 94–97, 119–123, 128–131; cycles 105–106, 119–120, 125, 71–72, 83–85, 117–121, 30–33, 72–75, 94–97, 140–142, 97–103, 142–143, 167–169; flight 105–106, 121, 71–75, 110–111; goal testing 32–33; inertia & friction 73–74, 94–96, 111–112, 72–73; limits or stopping 109–110, 119–120, 17–18, 71–72, 85–86, 109–111, 117–121, 48–50, 73–74, 128–131; map 106–111, 38–48; patterns 79–82, 92–94, 146–147, 11–31, 48–50, 52–55, 63–66, 117–121, 114–117, 142–143; reversing 109–111, 73–74, 169–172; states 126–129, 109–112, 117–121, 32–33, 75–79, 82–88, 93–103, 142–143
multiple choice 43–45, 51–62
multiple condition handling 115–116, 61–62, 101–103

music: about 12–13, 24, 56–57, 61, 65, 70–71, 77, 80–81, 83–84, 136, 2, 12–13, 64–65, 146–153; looping 71–72, 64–65; projects 71–72, 28–68
mute 83–84
my blocks 42, 47, 58–61, 81–83, 91–94, 134–135, 9–11, 34–35, 73–75, 94–101, 128–131

N

narrative games 32–68, 28–68; *see also* paths
nested ifs 8–10, 32, 43–45, 109–111, 115–116, 123–124, 32–33, 38–52, 58–61, 74–75, 82–84, 91–97, 112–114, 119–123, 142–145, 154–158
nested loops 82–83, 139–141, 11, 20–23, 26–29, 144–145
nesting 126–129, 139–141, 43–45, 80–81, 83–85, 21–22, 142–143, 154–158
not logic 43–45, 109–112, 148–150, 36–38

O

object property referencing *see* referencing sprite properties
obstacles 88–91, 93–94; *see also* hazards
offline scratch 4, 20–21, 27, 29, 164–165, 158, 169–170, 193–195, 206
off-screen deletion 71–72, 112–114
off-screen protection 37–38, 17–18, 104–106, 135–136, 17–19, 125–127
operators 42, 96–97, 106–108, 120–124, 127–131, 158, 160–161, 4, 18–19, 23, 26–29, 36–39, 43–45, 71–75, 81–91, 94–98, 106–111, 113–121, 123–124, 129–132, 139–140, 148–150, 165–166, 14–23, 32–62, 73–105, 110–158, 198–204; *see also* and logic; not logic; or logic
opponents 123–124, 76–78, 83–85, 114–117, 123–125, 128–145, 175–176
or logic 161, 36–38, 138–140, 166, 73–75, 91–92, 94–97, 117–119, 131–137, 201–204

order of operations 38, 47, 139–141, 146–147, 160–161, 36–37, 91–94, 112–113, 134–135, 140, 148–150, 166, 21–22, 32–35, 38–48, 50–62, 73–79, 82–84, 94–101, 117–123, 135–137, 142–143, 154–158, 201–204

outline (art) 51, 54, 87–89, 48–52, 74–75, 88–91, 104–106, 112–113, 17–19, 75–79, 114–117, 119–123, 125–131

over-size 107, 104–106, 109–111, 17–19, 75–79, 125–127

P

paint, custom sprites 51, 62, 89–91, 106–108, 112–114, 137, 24, 26, 41–43, 48–50, 74–75, 80, 88–91, 98, 100, 104–107, 116–119, 121, 124, 128, 16–17, 36–38, 58–61, 71–72, 79–90, 93–107, 110–117, 119–123, 125–140, 146–161

paintbrush *see* brush tool

paint bucket *see* fill

parameter 7–8, 94–97

parent/child concept 123–124, 163–164, 71–72, 74–77, 143–144, 168, 16–17, 97–101, 108–109, 128–131, 135–137, 165–167, 172–176, 204–205

paths, branching 39–40, 43–45

paths, merging/converging 51–53

pen tool 11–31

perspective 106–109, 35–36, 71–72, 163–165

physics *see* gravity; inertia & friction

pixel size 35, 107–108, 26–27, 32–33, 125–127, 163–165

plagiarism 28, 32, 62, 8–10

platform (computing) 17, 24–26, 153–154, 135–136, 146–148, 169, 171–172, 1–3, 7–13, 206, 208–210

platforms, terrain types 93–94, 97–103

play sound *vs.* start sound 159–160, 164, 198–201

playtesting 37, 169–170, 48–50, 112–117, 181–182

points (score) *see* scores

points (vector art) 48, 50, 53, 87–89, 74–75, 98–100, 104–106, 128–131

popups, making 28–29, 36–38, 52–61, 163–165, 169–172

positioning *see* glide; Go To; X/Y

position limits *see* off-screen deletion; workarounds, offscreen movement

position testing 109–110, 119–120, 125, 71–72, 32–33, 73–74, 82–84, 112–117, 125–127, 142–143

power ups 96–98, 123–124, 93–94

predicting mistakes 154–164, 36–37, 159–169, 179–180, 196, 198–205

procedural generation 13, 108–109, 21–29, 87–91, 96–98, 108–109, 114–117, 123–125, 135–137, 162–165, 175–176

progression systems 35–36, 80–83, 112–113, 17–19, 28–68, 38–44, 50, 61–62, 89–92, 123–125, 138–140, 162, 172–175

projectiles 71–72, 83–85, 112–114, 119–123, 128–131, 135–137

project page 28, 32, 94–95

projects 8, 13–14, 23, 25–28, 35, 57–58, 144–145, 150–151, 165, 167–168, 1–5, 8–10, 133–134, 140, 144–156, 158, 169, 171–172, 1–11, 162, 176–186, 192–195, 206

properties 54, 58–59, 63–64, 79, 80–81, 97–98, 137–139, 156, 163, 8–10, 13–14, 15–16, 103, 105–109, 135–136, 138, 144, 160–164, 4–8, 108–109, 135–137, 165–167, 175–176, 182–183, 196–198, 198–201

pseudocode 165–166, 170, 9–11, 206–207

R

radial gradient 54, 91, 88–91, 114–117, 119–123, 128–131

random generation *see* procedural generation

random intervals 87–91

randomization: chance events 27–29, 123–125; data/variables 23, 14–16;

direction/movement 119–120, 123–124, 36–37, 83–85, 142–143; play dynamics 158–161, 175–176; position 108–109, 121–124, 25, 87–91, 96–98, 131–134
rate of fire (ROF) *see* cooldown
reactions 52–57, 83–85, 50–61
reading code 146–147, 149–150, 160–161, 148–150, 152, 159–160, 182–184, 196
readouts *see* variable display
rectangle tool 52–53, 87–88, 107, 112, 23–29, 43–45, 48–50, 100, 104–107, 116–117, 124–127, 16–19, 71–72, 75–79, 85–88, 97–103, 112–114, 125–131, 135–137
recursion 20–21, 36–37, 43–45
referencing sprite properties 106–109, 138, 22–23, 55–58, 97–101, 114–117, 128–131, 135–137, 165–169, 182–183
relative positions *see* referencing sprite properties
remixing 26–28, 32, 148, 153, 8–10, 151, 158, 1–3, 69–70, 193–195; *see also* inside
repeat loops: about 82–83, 92–93, 131, 140, 146–147, 94–97; animation fx 78, 96–97, 99–100, 41–42, 48–50, 80–81, 121–123, 85–88, 119–123, 144–150; animation movement 78–81, 92–94, 104–108, 15–29, 36–37, 48–50, 58–61, 71–72, 94–101, 125–127; cloning 153, 96–98, 158, 17–19, 123–125, 193–195; data processing 26–29, 17–19, 21–22; nested 82–83, 20–23, 26–29; pattern making 15–29; randomizing 26–29, 88–91, 14–16, 123–125, 146–150; timing mechanism 80–81, 92–93, 106–108, 17, 48–50, 97–101, 125–131; until clause 71–72, 112–113, 134–135, 32–33
replay 94–95, 105–106, 109–110, 112–116, 91–94, 104–106, 116–117, 124–127, 69–70, 85–90, 110–111, 146–153

reset 94–100, 123, 14–15, 29–31, 34–35, 48–50, 106–107, 113–115, 105–107, 110–111, 146–153
reshape tool 51, 53, 87–89, 139, 74–75, 98, 104, 117–119, 129–132, 146–148, 128–131
resize 48–50, 52, 87–89, 112–113, 150, 26–29, 62, 78, 104–105, 111–112, 152, 14–16, 35–36, 52–55, 128–131, 185–186
restart *see* replay; reset
reticle 49, 107, 74–78, 16–17, 23–26, 75–79
reverse movement 109–111, 117–121, 73–74, 94–97
rgb *see* colour
room-specific sprites *see* scene-specific actions
rooms *see* scenes
rotation 49–52, 63, 96–97, 142–143, 24, 34–35, 104–107, 30–31, 72–73, 75–79; *see also* direction
run once protection 45–48, 45–48, 50–61, 82–84, 89–92, 138–140

S

saving data/progress 38–39, 112–115, 129–132, 140–141, 89–92, 138–140, 172–175
saving projects 26–27, 31, 34, 57–58, 154, 164–165, 158, 169, 193–195, 206
say/think blocks 129–131, 159, 163, 32–68, 164, 168, 48–61, 94–97, 198–201, 204–205
scaling difficulty 123–124, 158–161
scaling, perspective 28–29, 35–36, 163–165
scene changing *see* transitions
scenes 35–36, 39–45, 104–106, 129–132, 38–48, 172–175; *see also* transitions
scene-specific actions 45–48, 51–57, 81–83, 115–123
scores *see* variables, scores
screen refresh *see* frame rate, my blocks

scripts *see* stacks
see inside 28, 94–95, 144
selecting a backdrop 59–60, 70, 87, 136–137, 39–43, 45, 96–98, 140–141, 148–150, 28–31, 36–44, 48–50, 65–68
select tool 51–52, 88, 91, 107, 48–50, 62–63, 80–81, 98–100, 117–121, 64–65, 82–84, 119–123
sensing blocks 42, 109–110, 15–16, 35–36, 76–78, 85–91, 106–109, 111–112
sensory issues 21, 164–165, 169–170, 206
sequences: animations 104–105, 121, 41–43, 48–55, 58–68, 121–123, 34–35, 65–68, 85–88, 110–111, 119–123, 128–131, 146–150, 163–165; code block connections 38–39, 41–47, 139–140; drawing routine 11–31
sfx *see* sound effects; special effects
sharing 25–29, 31–32, 35, 154
show *see* visibility
simulation 12–14, 22, 69–101, 102–132, 169–172, 176–179
size: animating 95–97, 48–50, 146–150; art 35, 48–49, 87–91, 107, 112, 26, 16–17, 125–127; limits, workaround 107, 104–106, 135–136, 16–17, 125–127; property 63, 104, 135, 13, 71, 76, 78, 104–106, 135–136, 72–73
social learning 13, 154–155, 169–170, 150, 159–160, 9–11, 193–207
sound, volume 83–84, 125–127, 138–140, 146–150
sound editing 56, 64–65
sound effects 77, 80–81, 98–99, 131–133, 58–61, 112–114, 64–65, 117–123, 128–131, 138–140, 144–150; *see also* music
sound issues 21, 159–160, 164, 125–127, 198–201
sounds tab 56, 61, 65, 71–72, 77, 98–99, 131–133, 135–136, 64–65, 117–123
special effects 99–100, 29–31, 41–43, 58–66, 78–81, 119–123, 128–131, 138–140, 144–150; *see also* graphical effects

sprite library 58, 61–62, 65, 72, 89–90, 104, 119, 123, 34–35, 39–40, 45–53, 55–57, 58–61, 63–66, 71–72, 76–80, 96–98, 124–127, 129–132, 30–31, 38–44, 50–61, 65–68, 72–73, 75–79, 114–117, 151–161
sprites 38, 40–41, 58, 60–65, 72, 75, 89–91, 104, 106, 112, 114, 119, 123, 136–137, 155–157, 13–16, 23–26, 26–29, 35–36, 58–63, 104–106, 135–136, 138, 143–144, 160–162, 167–169, 22–23, 75–79, 114–117, 135–137, 165–167, 175–176, 196–198, 204–205
stacks 42, 146, 150, 162, 60, 152, 165–167, 14–16, 110–111, 185–186, 201–204
stage: backdrops 58–60, 69–72, 87–89, 104, 119, 13–14, 23–26, 39–40, 104–106, 38–44, 82–84; code for 60–61, 70–72, 83–84, 104–109, 140–141, 14–23, 30–31, 48–50, 114–117, 123–125, 138–140, 158–161; as object 58–61, 70–72, 156, 24, 39–40, 140–141, 14–16; window 35, 36–37, 58–60, 70, 108, 120, 138, 13–14, 17–18, 23–26, 35–36, 87–88, 104–106, 111–112, 135–136, 30–31, 73–74, 82–84
start game *see* game start
start screen 112–114, 100–101, 142–143, 65–68, 105–107, 151–153
state machines: about 128–131, 141, 161–162, 115–116, 140–141, 171–172; character states 75–79, 82–84, 85–88; gameplay/pause 115–116, 117–121; levels 79–81, 91–92; lives & death 125–128, 82–88; menu/inventory 36–38, 151–161; player turns 81–83; rooms/scenes 38–48; story key/lock 45–48, 52–57, 28–29, 36–38, 50–62; timer locks 113–115, 112–114, 138–140; *see also* game states
step (movement) 38, 119–120, 125, 138, 13–23, 26–29, 48–50, 52–55, 71–74, 94–96, 109–111, 117–121, 112–117, 128–137

stop 37, 42, 44, 74, 109–110, 114–116, 125–129, 131, 48–50, 52–55, 63–68, 98–100, 63–64, 82–88, 104–105, 114–117, 119–123, 125–127, 138–142, 144–150
studios 26, 27
switches *see* state machines; key/lock systems

T

tabs *see* backdrop tab; code tab; costumes tab; sounds tab
terrain 111–112, 71–72, 93–97, 97–103, 169–172
terrain-dependant actions 74–75, 93–97, 167–169
testing text values 35–38, 43–45
text: coding 17, 148–150, 171–172, 1–3, 7–8, 14–16, 179, 208–210; as data 41, 43, 45, 36–38, 48–50; dialogue 129–131, 159, 35–36, 41–43, 45–55, 58–68, 87–88, 98–100, 164, 50–61, 63–64, 146–150, 198–201; input 35–38, 139; text, joining 129–130, 38–39, 41–43; text, shadow/highlight effect 112–113, 80–81, 98–101, 65–68; tool (graphics) 51, 53, 112–114, 48–50, 80–81, 98–101, 116–117, 121–123, 129–132, 23–26, 36–38, 105–107, 125–127, 146–150, 154–158
time lines 34–35, 97–101, 163–165
timer, limit 138, 97–101, 112–114, 128–131, 165–167
timer, stopwatch 112–117
timers 113–117, 138, 34–35, 97–103, 112–114, 128–131, 163–167, 175–176
timing, animation 72–73, 78, 80–83, 92–93, 107–109, 141–143, 158–160, 17, 39–43, 45–57, 80–81, 121–123, 164–167, 28–29, 32–35, 72–75, 94–103, 119–123, 146–150, 163–165, 198–204
title screen *see* start screen; menus
tools 51–54; *see also specific tool names*
touching *see* collisions

touching, limitations 162, 96–98, 167–168, 204–205
transitions: backdrop switch 39–40, 38–44, 61–62; boss battle 138–140; death restart 85–88; fade to black 41–43; level 79–81, 91–92; overlays 112–116, 81–83, 98–101, 116–117, 124–127, 36–44, 63–64, 104–107, 119–123, 125–127, 146–158; switching sides 45–48; title screen 112–114, 100–101, 124–127, 63–64, 105–107, 151–153
transparency, art 53–54, 89, 91, 112, 114, 26–29, 74–75, 104–106, 75–79, 112–117, 119–123, 125–127, 135–137
transparency, ghost fx 63, 99–100, 139, 163, 41–43, 48–50, 58–61, 66–68, 80–81, 168, 85–88, 101–103, 119–123, 144–145, 204–205
troubleshooting 146–147, 152, 157, 9–11, 179–180, 193–207
true *see also* if; until
turns 80–83

U

undo 34, 48, 56, 139, 109–111, 117–121, 74–75, 169–172
until 44, 46, 71–72, 126–127, 132–133, 158–160, 162, 71–72, 112–115, 134–135, 162–167, 32–33, 64–65, 138–140, 146–153, 198–204
uploading 22, 31, 34, 56, 61–62, 106–107, 125–127
user interface *see* controls; displays; variable displays

V

value (coding concept) 17, 44–47, 54, 139, 158–159, 161, 18–19, 23, 43–45, 72–73, 85–86, 106–109, 135–138, 163–165, 12–16, 21–26, 32–33, 74–75, 94–97, 181–183, 198–204
variable displays: for data 14–16, 23–26, 27; show & hide 98–101,

124–127, 129–132, 151–158; sliders 18–19, 21–23, 72–74; techniques 138, 142–143, 163–167, 172–176

variables: about 42, 46, 110–111, 122–123, 125–128, 136–138, 140–143, 19–21, 165–167, 172–176; health/lives 125–128, 82–88, 117–123; hiding/showing 38–39, 80–81, 98–101, 116–118, 124–127, 36–38, 151–158, 163–165; limiting 72–73, 86–87, 123–124, 138; player controls 18–19, 21–23, 72–74, 86–87, 106–109, 115–116, 72–74, 82–84, 167–169; randomizing 23, 26–29, 138, 14–16; scores 110–111, 122–123, 87–88, 113–115, 117–119; special types 12–13, 108–109, 154–158, 165–167; testing 122–123, 125–128, 45–48, 51–53, 112–113, 123–124, 21–22, 32–35, 50–62, 79–105, 123–125, 138–140, 142–143, 154–158, 172–175; and text 129–131, 32–38, 45–48, 56–58

vector art, using 50–51, 53, 54–55, 62, 87–91, 107–108, 112–115, 23–29, 48–50, 55–57, 62–63, 74–75, 80–81, 88–91, 98–101, 104–107, 112–113, 117–132, 16–19, 36–38, 65–68, 71–72, 75–107, 110–117, 119–123, 125–140, 146–161

victory conditions 87–88, 112–113, 116–118, 61–62, 104–105, 144–150

views 35, 40–41, 48–49, 90, 146, 104–106, 94–97, 167–169

visibility 63, 97–100, 110–111, 123–124, 139, 163, 23–29, 34–35, 39–40, 45–50, 52–57, 66–68, 74–77, 80–81, 88–91, 96–101, 112–118, 121–129, 135–136, 143–144, 167–168, 14–19, 36–44, 48–61, 63–68, 75–88, 93–107, 110–123, 128–131, 135–140, 146–161, 204–205

volume 41, 57, 83–84

W

wait 72–76, 80–81, 100–101, 141–142, 8–10, 21–23, 45–48, 76–78, 80–83, 87–88, 106–107, 112–129, 134–135, 138, 21–22, 34–35, 38–44, 55–56, 65–68, 79–88, 94–103, 105–107, 114–131, 135–137, 140–142, 144–145, 151–153, 158–161

walking 125, 36–37, 48–50, 30–31, 72–73

waypoints 112–113, 89–90, 172–175

win screen 66–68, 116–118, 63–64, 104–105, 146–150

workarounds: answers 38–39; clone-specific inheritance 135–137; drawing circles 17–18; maximum size 104–106; minimum size 17–19; mutual destruction 119–123; offscreen movement 125–127; run once collisions 45–48; touching options 96–98

workspace 36, 38–41, 48, 56–57, 64, 107, 139, 74–75

X

X/Y: changing 79–81, 92–93, 105–106, 108, 26–29, 58–61, 71–72, 78–80, 94–96, 32–33, 72–75, 79–81, 85–88, 94–101, 110–111, 114–117, 125–127, 131–134, 138–140, 142–143, 146–153; coordinate system 37–38, 41, 45, 49, 79, 112, 137–138, 41–43; glide to 36–37, 45–48, 52–55, 30–31, 55–61, 63–64, 85–88, 138–143, 146–150; position 49, 94–95, 71–72, 88–91, 19–21, 32–33, 35–36, 38–44, 55–58, 79–90, 112–117, 125–127, 128–137, 142–143, 163–165; setting 94–95, 26–29, 52–55, 114–117, 131–104; troubleshooting 156, 158–159, 161, 161–165, 196–204

Z

Zoom 40, 48, 90, 146

For Product Safety Concerns and Information please contact our EU
representative GPSR@taylorandfrancis.com
Taylor & Francis Verlag GmbH, Kaufingerstraße 24, 80331 München, Germany

www.ingramcontent.com/pod-product-compliance
Lightning Source LLC
Chambersburg PA
CBHW060511300426
44112CB00017B/2629